ENERGY HARVESTING

Edited by **Reccab Manyala**

Energy Harvesting
http://dx.doi.org/10.5772/intechopen.71093
Edited by Reccab Manyala

Contributors

Juris Blums, Ilgvars Gornevs, Galina Terlecka, Vilnis Jurkans, Ausma Vilumsone, Sang-Jae Kim, Nagamalleswara Rao Alluri, Arunkumar Chandrasekhar, Gordon Thomson, Daniil Yurchenko, Dimitry Val, Fuhong Dai, Diankun Pan, Reccab Ochieng Manyala

First published in London, United Kingdom, 2018 by IntechOpen
IntechOpen is the global imprint of INTECHOPEN LIMITED, registered in England and Wales, registration number: 11086078, The Shard, 25th floor, 32 London Bridge Street
London, SE19SG – United Kingdom
Printed in Croatia

British Library Cataloguing-in-Publication Data
A catalogue record for this book is available from the British Library

Additional hard copies can be obtained from orders@intechopen.com

Energy Harvesting, Edited by Reccab Manyala
p. cm.
Print ISBN 978-1-78923-392-6
Online ISBN 978-1-78923-393-3

Meet the editor

Reccab Manyala is currently an Associate Professor of Physics at the University of Zambia where he has given lectures on Physics and carries out research and consultancy in New and Renewable Energy and Renewable Energy Technologies. He obtained his Ph.D. degree in Renewable Energy and Renewable Energy Technologies from the Maseno University (Kenya) in 2009. He holds his Master's degree in Theoretical Physics (Plasma Physics and Mathematical Physics) from the University of Nairobi (Kenya) and in Experimental Physics (Condensed Matter Physics) from the University of Arkansas (USA). He has published extensively and has research and teaching experience in many areas of physics spanning over 30 years in Kenya and abroad.

Contents

Preface IX

Chapter 1 **Introductory Chapter: Trends in Research on Energy Harvesting Technology 1**
Reccab Ochieng Manyala

Chapter 2 **Hybrid Structures for Piezoelectric Nanogenerators: Fabrication Methods, Energy Generation, and Self-Powered Applications 5**
Nagamalleswara Rao Alluri, Arunkumar Chanderashkear and Sang-Jae Kim

Chapter 3 **Wearable Human Motion and Heat Energy Harvesting System with Power Management 21**
Juris Blums, Ilgvars Gornevs, Galina Terlecka, Vilnis Jurkans and Ausma Vilumsone

Chapter 4 **Dielectric Elastomers for Energy Harvesting 41**
Gordon Thomson, Daniil Yurchenko and Dimitri V. Val

Chapter 5 **Piezoelectric Energy Harvesting Based on Bi-Stable Composite Laminate 63**
Fuhong Dai and Diankun Pan

Preface

From time immemorial, energy has held a pivotal position in any global, continental, regional or local economy. Many life-changing events have and are still taking place because of energy. Not only have economies flourished because of energy, some have completely collapsed. There is always a race toward the improvement of existing energy sources and supply or trying to find sustainable sources to address different problems such as global warming, new technology or just the knowledge that the world is running out of some forms of energy.

This book as a project, therefore, has been an attempt to bring to the fore ideas and information to those with interest in energy harvesting (EH) either for mere knowledge or for mainstream research of the methods and techniques that can be used to harvest energy for various applications.

The book has been organized in chapters where the introduction chapter gives a general view of global energy state, demand and some of the historical events, which have led to the current global energy situation. A brief review of the trends in energy research, motivation and direction is then presented. Finally, the main theme of the book (Energy Harvesting) is summarized by outlining the subject of the book chapter by chapter.

The first main chapter discusses smart energy harvesting through the surrounding environment, which can generate sufficient energy to drive low-power systems. This is seen as the forthcoming revolution in smart (or self-powered) technology, which may result in the phasing out of the usage of complex batteries and external circuit components and lead to the reduction in the depletion and the use of natural sources to generate energy for low-power systems. This chapter presents the fabrication methods, the growth of ZnO nanostructures on plastic substrates, flexible piezoelectric polymer film-based devices and how they were tested to improve the performance of piezoelectric nanogenerator (PNG) as a prominent energy harnessing approach for the development of sustainable independent power sources. This chapter concludes by stating that PNG technology suffering from brittleness, leakage current issues, high-electrical output generation and long-term durability can be possibly controlled by composite technology, i.e., polymer/nanoparticles.

Chapter 2 presents ideas on wearable human motion and heat energy harvesting system with power management options. This chapter starts by explaining that the combined human motion and heat energy harvesting system have been investigated. Types of the main parts of the developed human motion energy harvester have been described and voltage pulses in harvesters of different configurations have been explained including their designs and most importantly some energy harvesting methods and techniques such as thermoelectric generation. This chapter discusses in detail thermoelectric generator that harvests electricity from waste heat of the human body and concludes with a comparison of generated voltages and power at different activity levels and ambient temperatures.

The challenges and trends in energy harvesting through the use of radio frequency (RF) energy harvesting (EH) antennas are presented in Chapter 3. It is observed that the main challenge in implementing sensor devices for Internet of Things (IoTs) is the means for the operating power supply. The presentation then outlines how energy harvesting (EH) presents a promising solution as RF power is a suitable choice, particularly for cases where solar harvesting is not feasible. A number of challenges poised for the implementation of RF EH systems, especially for harvesting the ambient RF, are enumerated and discussed in detail. An exhaustive survey on RF EH systems is carried out and discussed. Important design issues are identified with insights drawn. In conclusion, the challenges in designing antennas suitable for harvesting RF power have been investigated and a general framework for designs suitable for ambient RF EH systems has been deduced.

In the last chapter, Chapter 4 of this book, the work on "Piezoelectric Energy Harvesting Based on Bistable Composite Laminate" is presented. The work discusses energy harvesting employing nonlinear systems and enumerates their advantages compared to linear systems in the field of broadband energy harvesting. This chapter also discusses how bistable energy harvesters have proven to be good candidates for broadband frequency harvesting due to their highly geometrically nonlinear response during vibrations. The work states that a potential category of bistable structures is bistable composite laminate, which has two stable equilibrium states resulting from mismatch in thermal expansion coefficients between plies. In conclusion, this chapter has summarized and reviewed the various approaches in piezoelectric energy harvesting based on bistable composite laminates.

Though the book covers a number of areas of energy harvesting, it is not exhaustive in the area of energy harvesting; however, the references and other sources given in the chapters will be very useful to anyone who would like to pursue more reading and research in energy harvesting.

I would like to gratefully thank Ivana Glavić, who acted as the Author Service Manager for the book. She worked tirelessly from the inception of the book idea to its final conclusion. Without her dedication, this project could not have come to fruition. I would also like to extend my sincere thanks to all the language copyeditors and all the technical team who spent many of their valuable hours to see the success of this project. Last but not least, my deepest thanks and appreciation are extended to Intech Open publisher who gave me the chance to edit another book on *Energy Harvesting*.

Reccab Manyala
University of Zambia
Department of Physics
The University of Zambia, Zambia

Chapter 1

Introductory Chapter: Trends in Research on Energy Harvesting Technology

Reccab Ochieng Manyala

Additional information is available at the end of the chapter

http://dx.doi.org/10.5772/intechopen.77314

1. Introduction

This introductory chapter gives a general view of global energy state, demand, and some of the historical events which have led to the current global energy situation. A brief review of the trends in energy research, the motivation, and direction are then presented.

2. Global energy status and research trends

The world demand for energy is projected to continue due to increased population and associated economic activities in addition to technological developments and changes. Environmental concerns related to traditional energy sources such as fossil fuels and recognition, that this is a diminishing resources globally, has given impetus in the development and search for a number of alternative energy sources. To date, new methods and techniques that allow energy to be harvested mechanically, electronically, magnetically, thermally, biochemically, or through some other means for different purposes and applications have been developed. However, some of the methods and techniques still require refinement and perfection in order to realize their full potentials. For example, despite solar cell technology having matured over the years, its full utilization of photovoltaic (PV) systems has been hampered by high costs related to processing technologies of silicon as well as limited conversion efficiency. In this respect, new materials are being tested to solve the high-cost problem.

Prior to the development of coal in the mid-nineteenth century, nearly all energy used was renewable. Energy that can be collected and naturally replenished on human timescale such as sunlight, wind, rain, tides, waves, and geothermal heat are termed renewable energy [1]. Four important areas—electricity generation, air and water heating/cooling, transportation, and rural (off-grid) energy—are the areas in which renewable energy provides services [2]. There

is no doubt that traditional biomass used for fuel fires is one of the oldest forms of renewable energy. The use of biomass has been traced back to about 790,000 years ago. According to much literature, biomass for fire did not become commonplace until many hundreds of thousands of years later, sometime between 200,000 and 400,000 years ago. Many authors put the second oldest usage of renewable energy as harnessing the wind in order to drive ships over water. The practice can be traced to ships in the Persian Gulf and on the Nile some 7000 years ago. However, historical records show that the primary sources of traditional renewable energy were human labor, animal power, water power, wind, in grain crushing windmills, and firewood, a traditional biomass.

Fears that civilization would run out of fossil fuels started to emerge in the 1860s and 1870s and the need was felt for a better source of energy.

The start of a major global energy crisis of the twentieth century began to unfold as petroleum production in the United States and some other parts of the world peaked in the late 1960s and early 1970s. World oil production in different oil producing countries began a long-term decline after 1979. The problems brought about by these events created a crisis which came to be known as the 1970s energy crisis. Major industrial countries of the world, particularly the United States, Canada, Western Europe, Japan, Australia, and New Zealand, faced substantial petroleum shortages, real and perceived, as well as elevated prices. The year 1973 saw the first worst oil crisis period. This was followed by another energy crisis in 1979, when the Yom Kippur War and the Iranian Revolution triggered interruptions in Middle Eastern oil exports [3–6]. The major industrial centers of the world were forced to contend with escalating issues related to petroleum supply. Western countries started creating reliance on the resources of potentially unfriendly countries in the Middle East and other parts of the world.

What followed after the 1970 oil crisis led to stagnant economic growth in many countries as oil prices rose sharply. Although there were genuine concerns with supply, part of the run-up in prices resulted from the perception of a crisis. The combination of stagnant growth and price inflation during this era led to the coinage of the term *stagflation*.

Both the recessions of the 1970s and adjustments in local economies saw many nations and countries finding ways to use petroleum in a more efficient manner. Strategies were developed and put in place which saw the petroleum prices worldwide return to more sustainable levels in the 1980s [7].

The 1970 oil crisis period was not uniformly negative for all economies. Petroleum-rich countries in the Middle East benefited from increased prices and the slowing production in other areas of the world. Some other countries, such as Norway, Mexico, and Venezuela, benefited as well. Major economic booms were experienced in the United States in places such as Texas and Alaska and some other oil-producing areas due to high oil prices even though most of the rest of the country struggled with poor and stagnant economy. Many of the economic gains, however, did not continue as oil prices stabilized and dropped in the 1980s [8–10].

While all these events were taking place, researchers worldwide intensified their efforts toward the search and development of energy that is collected from renewable resources, which are naturally replenished on a human timescale, such as sunlight, wind, rain, tides, waves, and geothermal heat for industrial and domestic uses. The sources called renewable energy sources often provide energy in four important areas: electricity generation, air and water heating/ cooling, transportation, and rural (off-grid) energy services.

With the emergence of new technologies employing semiconductors requiring low current and the realization that fossil fuels cause climate change and global warming coupled with high oil prices, the search for new and renewable energy and energy sources intensified. There has been increasing different world government support in terms of driving renewable energy legislation, incentives, and commercialization. New government spending, regulations, and policies that were implemented helped the energy industry weather the global financial crisis better than many other sectors. In a report of 2011 by the International Energy Agency, it is projected that solar power generators may produce most of the world's electricity within 50 years, reducing the emissions of greenhouse gases that harm the environment [11]. These sentiments have been echoed by the United Nations Secretary Ban Ki-moon.

As of 2011, small solar PV systems provided electricity to a few million households, and micro-hydro configured into mini-grids served many more. Though rural folks in least developed countries have no electricity, many households now rely on renewable energy such as biogas made in household-scale digesters for lighting and/or cooking. Others rely on new generations of more efficient biomass cook stoves. These efforts are, however, doing very little to address the bigger problem of global warming. On the other hand, the United Nations is highly in support of the renewable energy initiatives and believes that renewable energy has the ability to lift the poorest nations to new levels of prosperity [12].

Apart from the mainstream technologies using large amounts of energy, new gadgets in many households globally presently rely on continuous supply of small amounts of energy. Different forms of renewable energy sources are now in abundant supply for their operations. Mobile phones, portable Compact Disc players, and sport lights have evolved so much that most of them use rechargeable batteries. Though some of these gadgets are working quite well, researchers are still pushing the limits to perfect their energy supplies.

Author details

Reccab Ochieng Manyala

Address all correspondence to: reccabo@yahoo.com

Department of Physics, School of Natural Sciences, The University of Zambia, Zambia

References

[1] Ellabban O, Abu-Rub H, Blaabjerg F. Renewable energy resources: Current status, future prospects and their enabling technology. Renewable and Sustainable Energy Reviews. 2014;**39**:748-764. DOI: 10.1016/j.rser.2014.07.113

[2] Ramli MAM, Twaha S, Al-Hamouz Z. Analyzing the potential and progress of distributed generation applications in Saudi Arabia: The case of solar and wind resources. Renewable Sustainable Energy Reviews. 2017;**70**:287-297. DOI: 10.1016/j.rser.2016.11.204

[3] Jordan PJ, Albert BP. The Arab Oil Weapon–A Threat to International Peace. The American Journal of International Law. 1974;**68**(3):410-439

[4] https://en.wikipedia.org/wiki/1970s_energy_crisis [Accessed: March 15, 2018]

[5] Robert BB, Lutz K. Oil and the Macroeconomy since the 1970s. Journal of Economic Perspectives. 2004;**18**(4):115-134

[6] Garvin CC Jr. The oil glut in perspective. Annual API Issue. Oil & Gas Journal. November 9, 1981. p. 151-152

[7] Kilian L, Vigfusson RJ. Nonlinearities In The Oil Price Output Relationship. Macroeconomic Dynamics. Cambridge University Press; 2011;**15**(S3):337-363

[8] World: Saudis Edge U.S. on Oil. Washington Post; January 3, 1980. p. D2

[9] Gately D. Lessons from the 1986 Oil Price Collapsey. In: Brookings Papers on Economic Activity. Vol. 2. New York, NY: New York University; 1986. p. 239

[10] Ferrier RW, Bamberg JH. The History of the British Petroleum Company. Cambridge, UK: Cambridge University Press; 1982. pp. 201-203. ISBN: 978-0-521-78515-0

[11] https://news.un.org/en/story/2016/01/520182-renewable-energy-limitless-and-will-last-forever-says-ban-global-debate [Accessed: April 14, 2018]

[12] https://www.un.org/sg/en/content/sg/statement/2016-01-17/secretary-generals-remarks-international-renewable-energy-agency [Accessed: April 14, 2018]

Chapter 2

Hybrid Structures for Piezoelectric Nanogenerators: Fabrication Methods, Energy Generation, and Self-Powered Applications

Nagamalleswara Rao Alluri,
Arunkumar Chanderashkear and Sang-Jae Kim

Additional information is available at the end of the chapter

http://dx.doi.org/10.5772/intechopen.74770

Abstract

Smart energy harvesting through the surrounding environment generates sufficient energy to drive the low-power consumption systems. It is the forthcoming revolution in smart (or self-powered) technology and results in abolishing the usage of complex batteries, external circuit components, and natural sources. To date, extensive fabrication methods, the growth of ZnO nanostructures on plastic substrates, and flexible piezoelectric polymer film-based devices were tested to improve the performance of piezoelectric nanogenerator (PNG) as a prominent energy-harnessing approach for the development of sustainable independent power sources. Still, PNG technology suffers from brittleness, leakage current issues, high electrical output generation, and long-term durability, which can be possible to control by the composite technology, that is, polymer/nanoparticles. The objective of this book chapter determines the rapid growth of multifunctional, flexible composite structures through various methods (e.g., ionotropic gelation method, groove technique, ultrasonication followed by solution-casting methods) for high output energy generation and self-powered sensor/system studies.

Keywords: piezoelectric nanogenerator, composite structures, self-powered sensors, $BaTiO_3$ nanoparticles, polyvinylidene fluoride (PVDF)

1. Introduction

The contemporary society and technology directing toward the digitalization of the world increases the huge consumption of electrical energy. Over the decades, the major percentage

amount of electrical energy is generated by the consumption of the natural sources like oil, gas, coal, and water. However, the natural sources are limited in supply, and the continuous energy generation using these sources creates global warming issues and pollution, which will effect on the lifespan of the human race. Moreover, the regeneration of the natural sources takes hundreds of the thousand years. The international energy agency (IEA) statistics depicts that the energy crisis is increasing day by day due to the lifestyle choices, climate change, industrialization of sectors, and typical regulation policies by power management sectors. The energy crisis arises due to two major problems: (1) the power supply and demand and (2) overconsumption of natural sources [1]. It is imminent to the present human generation and a serious issue for the future generation mankind. Alternative energy harvesting (AEH) options and energy policies are introduced across the world to overcome the energy crisis [2].

To drive any electronic device/sensor effectively depends on the following key factors such as interfacing circuit components, a processing unit, a sensing unit to measure/monitor various stimuli, and a battery source to power up the whole unit [3]. Among them, removing the toxic, complex wiring battery source and reducing the interfacing circuit components are the prime factors to improve the smart electronic device technology. This is possible only by the development of a nano/micro-sustainable independent power source coming from the surrounding environment. The AEH approaches will have many benefits such as saving natural resources, eco-friendly operation, portable/wearable, low cost, and finally removing the battery energy.

Many researchers believed and developed that the AEH approaches using the multifunctional nanomaterials and its devices will reduce the partial amount of energy crisis and make it possible to create a new type of multifunction smart sensors/devices in various fields such as biomedical, automotive, wireless-sensor networks, and communication electronics [4]. The rapid development, the invention of new nanomaterials, and cost-effective device designs are highly desirable to utilize the surrounding waste mechanical energy (machine vibrations, human body motions, wind/water motions, and ocean waves) in our surrounding environment [5].

Until now, piezoelectric nanogenerators (PNGs) [6], triboelectric electric nanogenerators (TNGs) [7], pyroelectric nanogenerators (PyNG) [8], and thermoelectric generator (TMNG) [9] are widely investigated and practically demonstrated to convert the waste mechanical/thermal energy into useful electrical energy. Among them, PNG technology is highly reliable, stable operations, smaller device area, and diverse application fields. To date, extensive fabrication methods, growth of various one-dimensional (1D)/two-dimensional (2D) inorganic piezoelectric nanostructures (NSs) on plastic substrates [6], flexible piezoelectric polymer films, and device designs (planar, stretchable, cylindrical, or fiber) were developed to improve the PNG technology as a prominent energy-harnessing approach for creating the sustainable independent power source to drive the low-power consumed electronic devices/sensors. Moreover, the device compatibility, electrical output performance (nW/cm^2 to $\mu W/cm^2$) under various harsh environments, and flexibility issues were optimized to think about the real-time commercialized PNG product [10]. On the other side, few PNGs have dual functionality such as wearable/portable independent power source to drive the commercial electronic devices and can also work as a self-powered sensor (or a battery-free sensor) to measure/monitor the various physical, chemical, biological, and optical stimuli [11]. The study reports suggest that PNG technology based on the type of materials is classified into three categories such as (1) inorganic nanostructures (NSs)-based PNG, (2) polymer matrix-based PNG, and (3) composite (polymer + inorganic NPs) PNG as shown in **Figure 1** [12]. Wang et al. reported the first PNG report-based ZnO nanostructures (named as inorganic PNG) and later showed the many possible ways to improve the instantaneous power density of the PNG [6]. Many other fellow researchers across the world extend the growth of other nanostructures and its device designs for inorganic PNG, but it has few ample drawbacks such as typical growth process of piezoelectric NSs, brittleness, lower force limits, failure instability, and leakage current issues [6, 10, 11]. Further, PNG was implemented with the flexible polyvinylidene fluoride (PVDF) and its copolymers due to its high flexibility, easy process to prepare flexible films, low electrical output performance, and accepting large mechanical force. However, it has one major disadvantage such as low piezoelectric coefficient and relative permittivity at room temperature than the inorganic

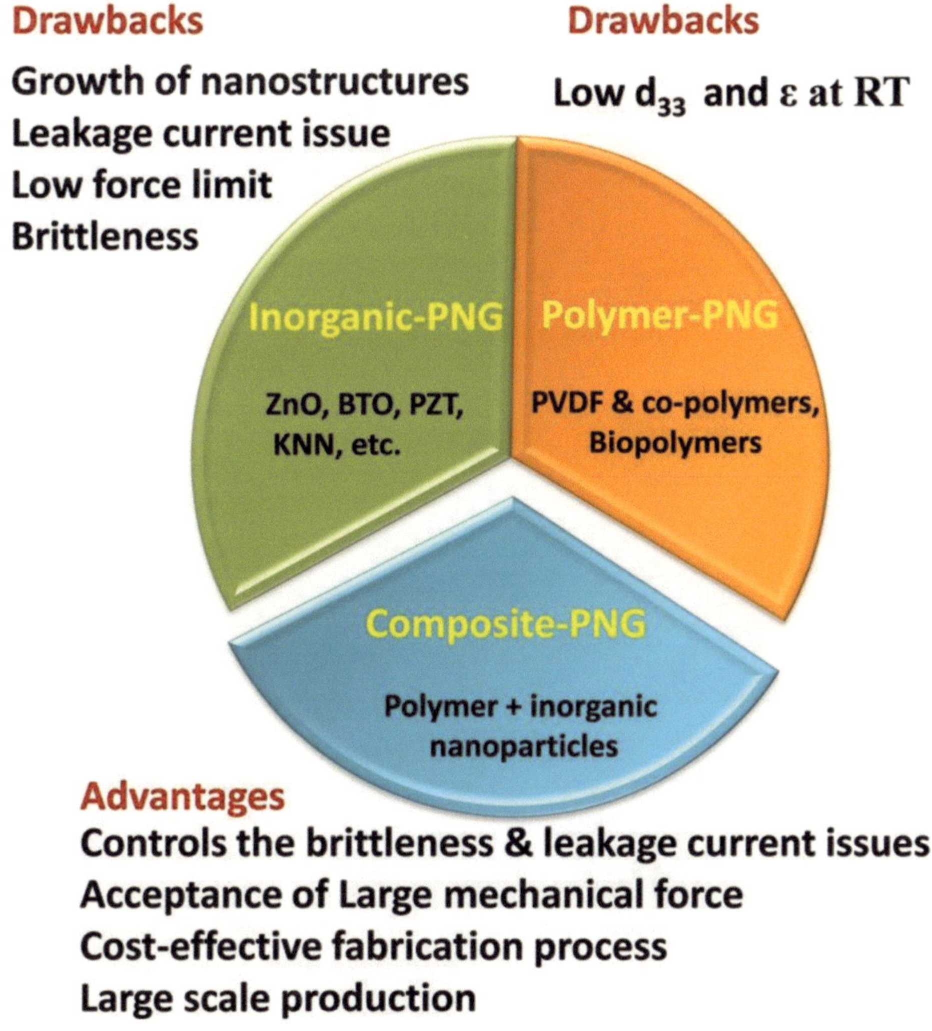

Figure 1. The schematic represents the overview of PNGs: classification, drawbacks, and advantages.

piezoelectric NSs. Recently, our group and many other research groups successfully overcome the issues of the inorganic and polymer PNGs by developing the composite PNG technology [13–15]. The major key factor in designing the efficient, flexible composite PNG device is the development of multifunctional hybrid or composite piezoelectric structures (HPSs). It indirectly depends on the selection of individual high-performance inorganic and polymer materials and cost-effective fabrications processes. These kinds of composite PNGs are highly suitable to work as sustainable independent power sources as well as self-powered sensors to measure various physical parameters such as physical, optical, biological, and chemical stimuli.

The chapter describes the cost-effective HPSs (flat, spherical beads, worm structures, micropillars, irregular films, and hemispherical strips) developed by the simple fabrication processes such as solution-casting technique (SCT) [13, 14], ionotropic gelation process (IGT) [15], and groove technique (GT) [16]. The digital photographs of the as-fabricated HPSs are shown in **Figure 2a**. The HPSs were fabricated using the particular weight ratio of the

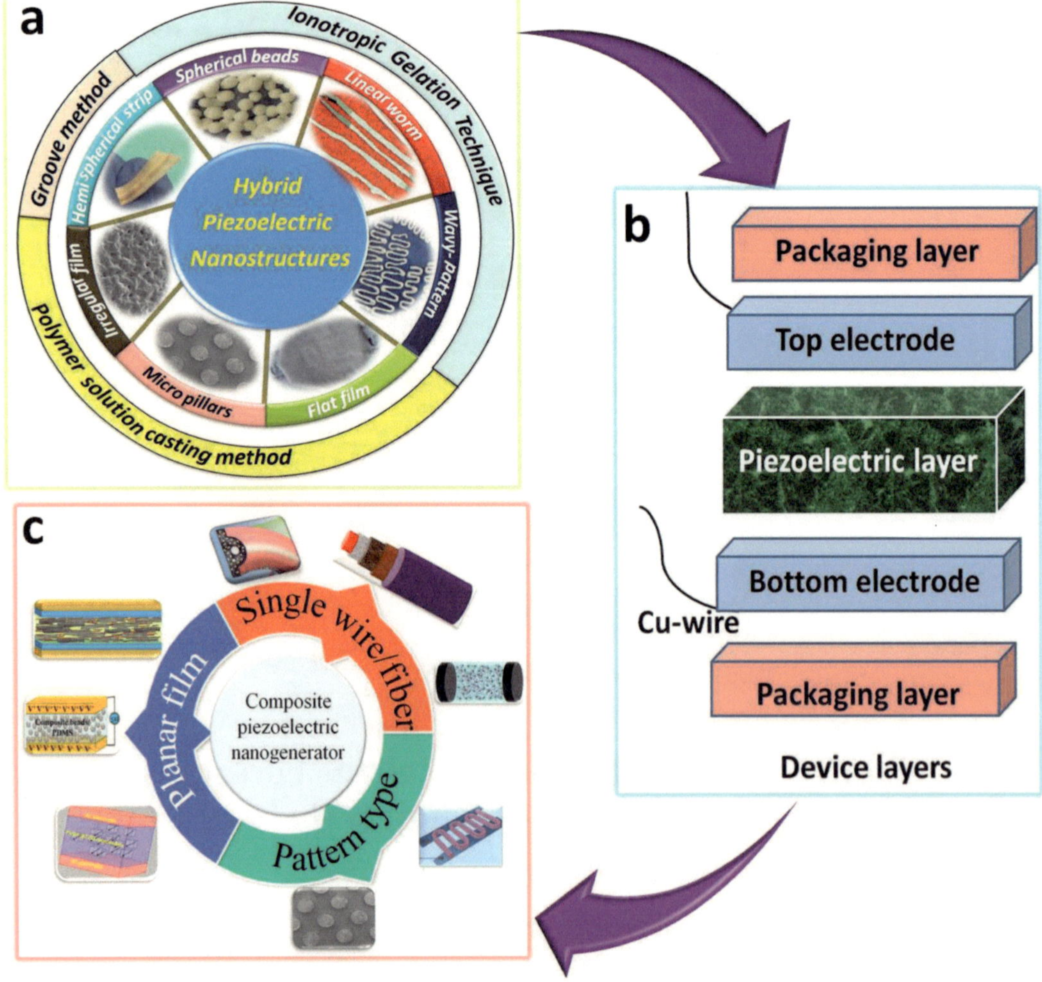

Figure 2. (a) The schematic represents the proposed HPSs using the polymer SCT, ITG, and GT. (b) Device layers of the composite PNG. (c) Possible types of composite PNGs such as planar film, pattern type, and single wire/fiber.

inorganic nanostructures such as $BaTiO_3$ nanoparticles and nanocubes, and the polymers are PVDF (polydimethylsiloxane), PDMS, and calcium alginate biopolymer. Further, these structures are used to fabricate the composite PNG devices. The device structures have three main layers such as (1) piezoelectric active layer (i.e., HPSs) and (2) top/bottom electrode layers (aluminum (Al), gold (Au), and platinum (Pt)) on the HPSs and top/bottom packaging layers (PDMS) as shown in **Figure 2b**. The sandwich structure of all these layers without an airgap forms as a composite PNG device. Based on our previous works, the composite PNGs classified as three types such as planar film, pattern type, and single wire/fiber-type PNGs, and the corresponding schematic diagrams are shown in **Figure 2c**. Further, these PNGs are used as self-powered sensors/systems to monitor the fluid velocity and pH value of the alkaline solution. Self-powered flexion sensor to classify/measure the individual finger motions, nonlinear muscle motions of human body parts, acceleration sensors to measure the various accelerations of the linear motor shaft, and driving the commercial light-emitting diodes (LEDs)/liquid crystal display (LCD).

2. Experimental section

The hybrid or composite piezoelectric structures (HPSs) were developed using the homogeneous mixture of the various inorganic nanostructures, fillers into the polymer matrix. The shapes of the HPSs mainly depended upon the selection of the template polymer matrix, fabrication method, and the type of raw materials used (particular amount). In this case, the developed HPSs (flat film, spherical beads, linear and nonlinear worms, and hemispherical strip) are having multifunctional properties, and the process is eco-friendly, cost-effective, and easy to make the large-scale production at a given time interval. The followed synthesis procedures for functional nanoparticles and the fabrication methods for HPSs are solid-state reaction (SSR) nanoparticles [15], molten salt method (MSM) nanocubes [13], solution-casting technique (SCT)-flat film [13], ionotropic gelation process (ITG) spherical beads [15], worm structures [17, 18], and groove technique (GT)hemispherical strip [16].

2.1. Synthesis of $BaTiO_3$ nanoparticles and $BaTi_{(1-x)}Zr_xO_3$ nanocubes

2.1.1. $BaTiO_3$ nanoparticles (BTO NPs)

The crystalline, randomly oriented BTO NPs were synthesized by SSR method [15], and the corresponding possible chemical reaction is.

$$BaCO_3\,(\text{solid}) + TiO_2\,(\text{solid}) \leftrightarrow BaTiO_3\,(\text{solid}) + CO_2\,(\text{gas}) \tag{1}$$

Here, the precursor materials such as barium carbonate and titanium dioxide were taken according to the atomic ratio of the final product and thoroughly grinded in the ethanol medium. The obtained homogeneous powder was placed in an alumina tube and fired at an elevated temperature of 1200°C for 2 h in an open tube furnace. After cooling down to the room temperature, the alumina tube was removed from the furnace and the final product collected and crushed into a fine powder form.

2.1.2. $BaTi_{(1-x)}Zr_xO_3$ nanocubes (BTZO NCs)

The MSM for synthesizing the inorganic perovskite nanostructures (NSs) [12] reduces the processing temperature and improves crystalline quality as compared to the NSs prepared by the SSR method. Here, the BTZO NCs are prepared by the homogeneous mixer of the reactant precursors ($BaCO_3$, TiO_2, ZrO_2), the eutectic products (molten salts are NaCl, KCl, and Na_2SO_4) and fired at a temperature of 750°C for 3 h in an open tube furnace. After natural cooling, the obtained product was washed several times to remove the existing unwanted salts precursors and finally heated at 60°C/overnight in a hot air oven. The final product was again crushed into a fine, smooth powder used to evaluate the structural, surface morphological, and functional characterization studies. In this case, the various amounts of zirconium (Zr) foreign atoms (x = 0, 0.05, 0.1, 0.15, and 0.2) were doped into the B-site of the parent BTO lattice and the corresponding structural, energy harvesting capability was investigated thoroughly [12].

2.2. Fabrication of hybrid piezoelectric flat film

Various types of the hybrid piezoelectric flat (HPF) films were developed by the mixing of the particular weight ratio or volume percentages of the inorganic NSs into the fixed amount of the template polymer matrix. The performance, functionality of the as-fabricated HPF film depends upon the optimized weight ratio, the thickness of the film, and the homogeneous distribution of the NPs into the template polymer matrix. Here, PVDF/BTO NCs, PVDF/BTZO NCs [12], PDMS/BTO NCs [19], and PVDF/activated carbon (AC)-based HPF films [20] were fabricated and tested to harness the mechanical energy and to sense/monitor the physical, chemical, and optical stimuli.

2.2.1. Fabrication of the PVDF/BTZO NCs, PDMS/BTO NCs HPF, and PVDF/activated carbon (AC) films

Highly flexible, cost-effective PVDF/BTZO and PVDF/BTO NCs HPF films [12] were fabricated using the simple ultrasonication method followed by the SCT. The HPF film was made up of the two key steps. One, the transparent, homogeneous PVDF solution was prepared by the sonication (approximately 1 h) of the PVDF powder (1 gm) into an 8-mL mixed solution obtained from the solvents such as N,N-dimethylmethanamide (DMF) (5 mL)/acetone (3 mL). During the sonication process, there is a possibility to generate the high temperature and the polymerization of the PVDF, the phase may be possible to vary. Therefore, a cold-water bath setup arranged around the glass vials contains the PVDF powder/solvents. A clear transparent solution was obtained when the PVDF powder completely dissolved in DMF/acetone solution. Second is the particular weight ratio (0.5 or 1, etc.) of the as-synthesized BTO NCs were weighted, poured into the transparent PVDF solution, and the ultrasonication process continued for one more hour. For every 15 min of the sonication process, small mechanical agitation was performed in glass vials using the fine glass rod. There is a high chance that the clustered BTO NCs will settle at the bottom of the glass vial, and the generated ultrasound energy from the microtip probe will not dissolve these clustered BTO NCs into the PVDF solution. A clear non-transparent, whitish-colored hybrid solution was obtained after the 90% dissolution of the BTO NCs into a polymerized PVDF solution. The obtained whitish hybrid solution was removed from the sonication setup and cooled down to room temperature in an air atmosphere. After that, the hybrid solution was poured into fixed glass Petri dish and heat treated at 60°C overnight. Highly flexible PVDF/BTO NCs HPF film was obtained when the solution completely dried. Further, the structural and functional properties of the as-fabricated films (desired dimension) were investigated carefully. The fabrication process of the remaining HPF films such as PVDF/Zr-doped BTO NCs [12], PDMS/BTO NCs [19], and PVDF/AC [20] was similar and is explained in detail in our previous published reports.

2.3. Fabrication of hybrid (BTO NPs/calcium alginate) piezoelectric spherical beads (HSB), linear and wavy-pattern worm (LW, WPW) structures

The HSB [15], HLW [17], and HWPW [18] structures were fabricated using the eco-friendly, cost-effective ITG process. In this case, nontoxic polysaccharide, three-dimensional scaffold-structured biopolymer was used as a template matrix to prepare the HPSs. The IGT process mainly consists of three major key steps. One, the preparation of the clear transparent sodium alginate (Na-alg) solution using the alginic acid sodium salt (2 g) obtained from the brown algae magnetically stirred (50°C for 2 h) in particular amount of double distilled water (100 mL). Second is the preparation of the non-transparent hybrid (Na-alg/BTO NPs) solution by the homogeneous mixing of the as-synthesized BTO NPs into the obtained transparent Na-alg solution. The hybrid solution was thoroughly stirred magnetically at the elevated temperature of 70°C for 2 h without the sign of BTO agglomeration. The third is the gelation, replacement of sodium atoms in a hybrid solution by the calcium atoms to form as the ionic gelation of the BTO NPs/calcium alginate (BTO NPs/Ca-alg). Here, the substitution of the Ca-atoms is possible, when the squeezed hybrid (Na-alg/BTO NPs) solution is poured into a 1 wt% of calcium chloride ($CaCl_2$) bath. The procedure mentioned above is common for three structures such as HSB, HLW, and HWPW, and the only difference is the instrument used to drop the hybrid solution into the $CaCl_2$ bath, free-fall height between the droplet position, and $CaCl_2$ solution. The hybrid solution-loaded pipette dropper or a perforated metal mesh is used for HSB structure, the plastic syringe used for HLW structure, and the freehand drawing of the hybrid solution-loaded pipette dropper used for HWPW. The pure (Ca-alg) spherical beads and HLW structures were prepared by following the same protocol, whereas the pure wavy-pattern worm structures are not possible to prepare due to the high volatility of the Na-alg solution and the drawn pattern completely overlapped each other. Further, the structural, optical, and functional properties of the as-fabricated HPSs were investigated carefully [15, 17, 18].

2.4. Fabrication of hybrid (BTO NPs/PDMS) piezoelectric hemispherical strip

The adaptable, highly flexible hybrid piezoelectric hemispherical strip (HPHS) is fabricated using the innovative, cost-effective GT. This kind of structure, approach is highly suitable to prepare the unconventional HPSs for measuring the nonlinear mechanical motions. Here, the shapes of the piezoelectric strip mainly depend on the type of the groove used during the fabrication time. The GT for HPHS mainly has four key steps. One, the preparation of the hemispherical groove on PET sheet and the placing of the flexible metal wire (titanium (Ti) or copper (Cu)) exactly at the center of the groove by using the two rigid supports at the ends of the PET groove. Second is the preparation of the hybrid (PDMS/BTO NPs) solution, pouring the required solution into that PET groove. Third is placing the loaded PET groove into a hot air oven and fired at 70°C for 30 min. Fourth is the peel-off dried hybrid strip having one rectangular side base and another hemispherical side groove (replica of the PET groove) [16]. The flexibility, twisting capability, structural, functional, and energy-harvesting capability were evaluated and tested to monitor the human body physical moments.

Further, as fabricated, all the HPSs were utilized to harness the mechanical energy by fabricating the flexible/non-flexible PNGs and showed the possible prototype designs to monitor/sense the applied input stimuli without using the additional battery source, circuit components.

3. Results and discussion

The piezoelectric BTO and BTZO NCs prepared by the low-temperature MSM and the corresponding FE-SEM images at 200 nm are shown in **Figure 3a** and **b**. The Raman spectra of NCs have the active vibration modes. They are [A1(TO), E(LO)] at 196 cm^{-1}, A1(TO) at 256 cm^{-1}, [B1(TO+LO), E] at 310 cm^{-1}, and [A1(TO), E] at 520 cm^{-1}, and A1(LO), E] at 720 cm^{-1}. The major peak at 310 cm^{-1} confirms the existence of the tetragonal phase of BTO having the possibility of a higher piezoelectric coefficient than all other phases of BTO. Also, the crystalline phase confirmation of BTZO NCs, HPF films, was confirmed by the XRD patterns, FT-IR techniques, and the information was given in our previous published report [13]. As shown in **Figure 2b**, the composite PNG was fabricated using the as-developed HPF film (2.5 cm × 2.5 cm), top/bottom electrodes and top/bottom packaging layers, respectively.

Initially, the mechanical force 11 N was applied exactly perpendicular to the PNG0 [PVDF/BTO NCs] and PNG1 [PVDF/BTZO NCs (x = 0.1)] devices, and the corresponding electrical responses are shown in **Figure 3e** and **f**. Here, the PNG1 shows a higher electrical response (11.9 V, 1.35 μA) than the PNG0 response (7.99 V, 1.01 μA) due to the Zr-doped BTO NCs. Here, the PVDF solution amount, device area, and applied force were kept constant. And the only difference is the substitution of a particular amount of Zr^{+4} into the Ti^{+4} site in the BTO lattice, which will be possible to enhance the piezoelectric coefficient than the pure BTO NCs. Here, the PVDF polymer has dual functionality such as supporting crosslinker to hold the BTO NCs to form a hybrid film and the small contribution to enhancing the generated electrical energy due to the small piezoelectric coefficient of PVDF (β-phase). Further, the PNG1 was tested to measure the various fluid velocities of the water at the outlet of pipe, and the corresponding electrical responses are shown in **Figure 4a**. It demonstrates that the increasing water velocities (31.43, 78.6, and 125.7 m/s) at the outlet pipe will be possible to generate different mechanical forces that act on the PNG1 and to respond with the increased piezoelectric potentials (self-powered fluid velocity sensor). It suggests that the PNG1 is a potential candidate to work as self-powered fluid velocity sensor without using any external battery energy and the additional circuit components.

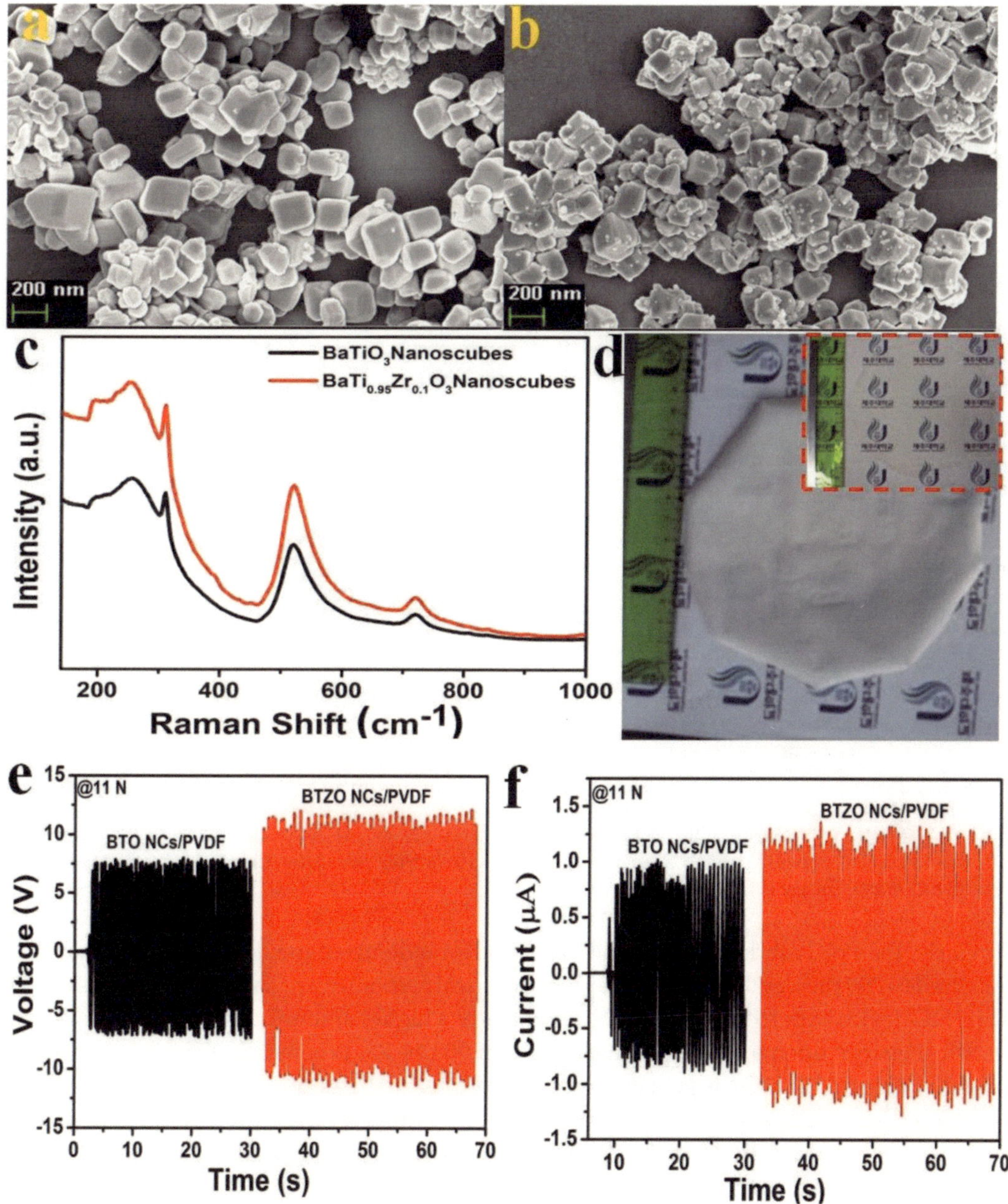

Figure 3. Flexible HPF film fabricated by solution-casting technique and its electrical response of PNGs. (a and b) High crystalline BTO and BTZO ($x = 0.1$) NCs synthesized by the MSM. (c) Raman spectra of the BTO and BTZO NCs. (d) Digital photograph of the as-fabricated non-transparent HPF film and the inset shows the as-fabricated transparent pure PVDF film. (e and f) Generated electrical response (voltage, current) of the PNG0 (BTO NCs/PVDF) and PNG1 (BTZO NCs/PVDF) upon mechanical force (11 N). The figures are reproduced with the permission from Ref. [13]. Copyright of American Chemical Society (ACS) publications.

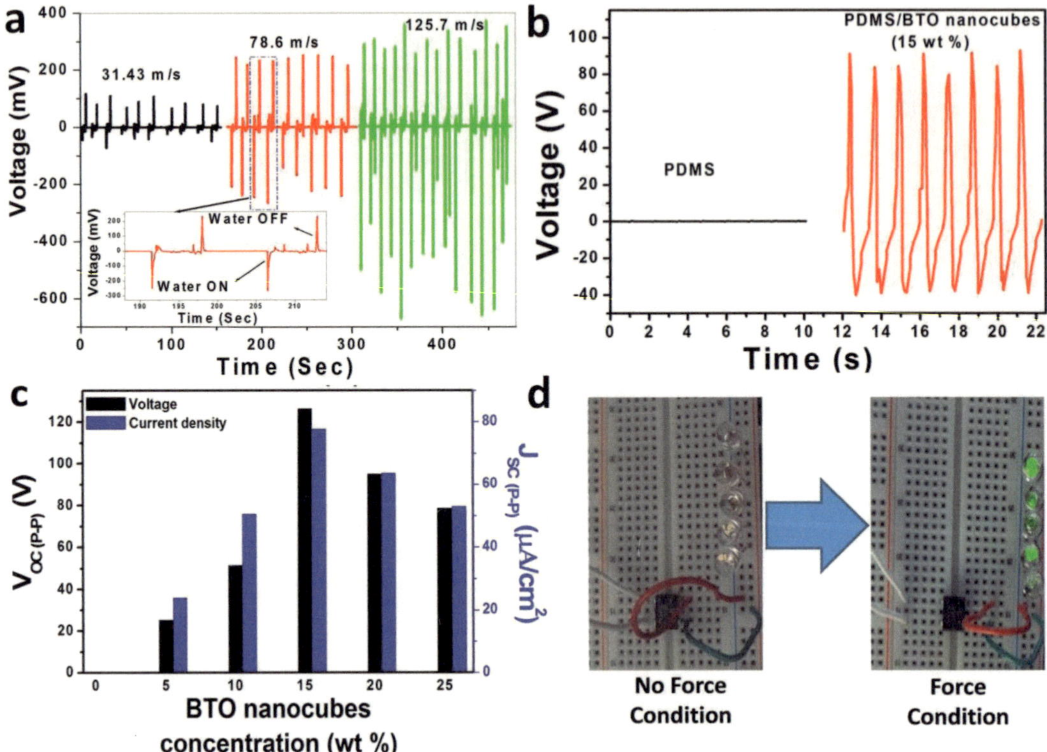

Figure 4. (a) Demonstration of the self-powered fluid velocity sensor using the PNG1. (b) Output voltage response of PDMS and PDMS/BTO NCs-based devices upon the mechanical force. (c) PNG device electrical responses as a function of the weight ratio of BTO NCs into the PDMS matrix. (d) Driving commercial five green LEDs using the PDMS/BTO NCs film-based PNG device (15 wt%). The figures are reproduced with the permission from Refs. [13, 19]. Copyright of American Chemical Society (ACS) publications.

Similarly, the PNG fabricated with the HPF film is made up of the flexible PDMS/BTO NCs (thickness ≈ 790 μm), and the corresponding electrical responses are shown in **Figure 4b** and **c**. The PNG device with 15 wt% of BTO NCs generates a higher amount of piezoelectric potential electrical energy upon a low mechanical force of 988.14 Pa. Whereas in the case of the PNG device without BTO NCs, it generates almost zero electrical output upon mechanical force, because PDMS matrix does not have any piezoelectric property (non-centrosymmetric behavior) as shown in **Figure 4b**. Also, the PNG electrical output upon constant mechanical force as a function of the weight ratio of BTO NCs into the PDMS has been verified, and the corresponding electrical response is shown in **Figure 4c**. It demonstrates that the linear increment output behavior is observed when increasing the weight ratio of BTO NCs (i.e., from 0 to 15 wt%) in PDMS matrix. Beyond the 15 wt% of BTO NCs into the PDMS, it will result in the decreased electrical output. It may be due to the agglomeration of BTO NCs in the PDMS results in the lower cumulative electric dipole moment, and at the same time, the applied mechanical force may not be possible to distribute evenly on the PNG surface. Next, the potentiality of the PNG device (15 wt% of BTO NCs) output was tested to drive the commercial purchased five green LEDs connected in series. LEDs were in OFF condition when no mechanical force acts on the

PNG and successfully lit up when the force acts on the PNG device as shown in **Figure 4d**. It demonstrates that the as-developed HPF film-based device is a highly potential candidate to work as a sustainable independent power source.

Researchers are directed toward the development of the cost-effective, eco-friendly fabrication methods to prepare the innovative HPSs-based PNG devices for various applications. For the first time, ionotropic gelation (ITG) technique and 3D template structure-based alginate biopolymer were introduced into the energy-harvesting field, and the possible HPSs for harnessing the waste mechanical energy is shown. **Figure 5** shows the digital photographs, the surface morphology of the as-fabricated HSB, HLW, and HWPW structures and the corresponding PNG device electrical responses. Here, the piezoelectric tetragonal crystalline phase of BTO NPs was not changed even after the substitution of BTO NPs into the Ca-alginate polymer. The HSB-based PNG device fabricated by mixing the particular amount of HSBs into the PDMS matrix forms a flat film, and the followed the above-demonstrated sandwich fabricating the structure. **Figure 5a(iii)** demonstrates that the higher electrical output voltage (≤11 V) generated for the HSB-based devices as compared to the pure PDMS (<3 V) and pure Ca-alg beads/PDMS (≤7 V)-based devices upon constant mechanical force. Here, a small amount of electrical

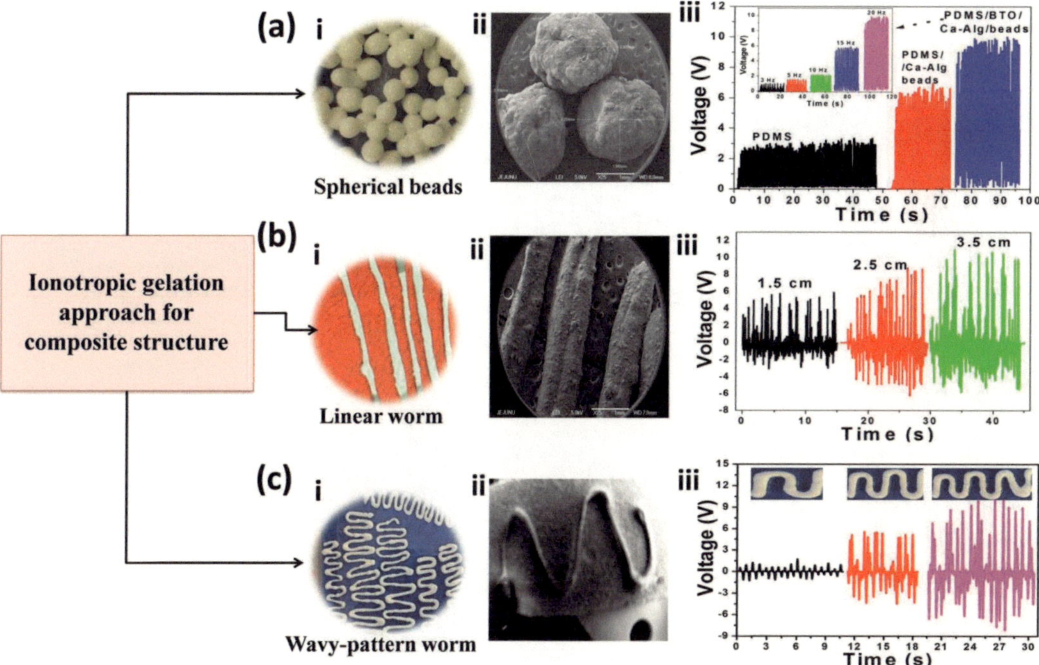

Figure 5. ITG approach for innovative HPSs and its device electrical responses: (a) digital photograph of HSB: (i) surface morphology and (ii) generated an electrical response of the PDMS-, PDMS/Ca-alg-, and PDMS/HSB-based devices upon the constant mechanical force. The inset shows the frequency-dependent electrical response of the PDMS/HSB-based device. (b and c) Digital photograph of HLW and HWPW structures: (i) surface morphologies and (ii) the length-dependent electrical response of the HLW and HWPW structure-based devices. The inset shows the various lengths of the optical photographs of the HWPW structures. The figures are reproduced with the permission from Refs. [15, 17, 18]. Copyright of Elsevier.

output voltage is generated when there is no BTO NCs in the beads and the PDMS matrix, due to the existence of the small air gap in the as-fabricated final device structure results in the introduction of the triboelectric effect of Ca-alg and PDMS layer along with the Al foil electrode. Therefore, the generated final electrical output of HSB-based PNG is due to the synergistic electric of the piezoelectric effect from the BTO NCs and triboelectric effect from the Ca-alg and PDMS layers. Further, the HLW and HWPW structure-based PNG devices were fabricated and tested to harness the mechanical energy. The generated HLW-based device positive output increased from (≈5 V) to (≈10 V) upon hand mechanical force, when the length of the HLW structure increased from 1.5 to 3.5 cm as shown in **Figure 5b(iii)**. The similar behavior is observed for the HWPW-based PNG devices, that is, the voltage increased from (≈1.5 V) to (≈10 V), when the length of the HWPW structure is from 0.85 to 1.95 cm as shown in **Figure 5c(iii)**. To identify the role of Ca-alg in these structures, the pure linear worm structure-based device (without an air gap) has been fabricated and the zero electrical output has been generated, which confirms that no triboelectric effect was introduced into these device structures. Further, these structure-based devices were used to measure the flexion/extension of the individual finger motions (human hand) and the cumulative fist movement, that is, self-powered flexion sensor (SFS) as shown in **Figure 6a** and **b**. Also, the monitoring of the pH value of the alkaline solutions ranges from 12 to 8 (self-powered pH sensor (SPS)) as shown in **Figure 6c** and **d**.

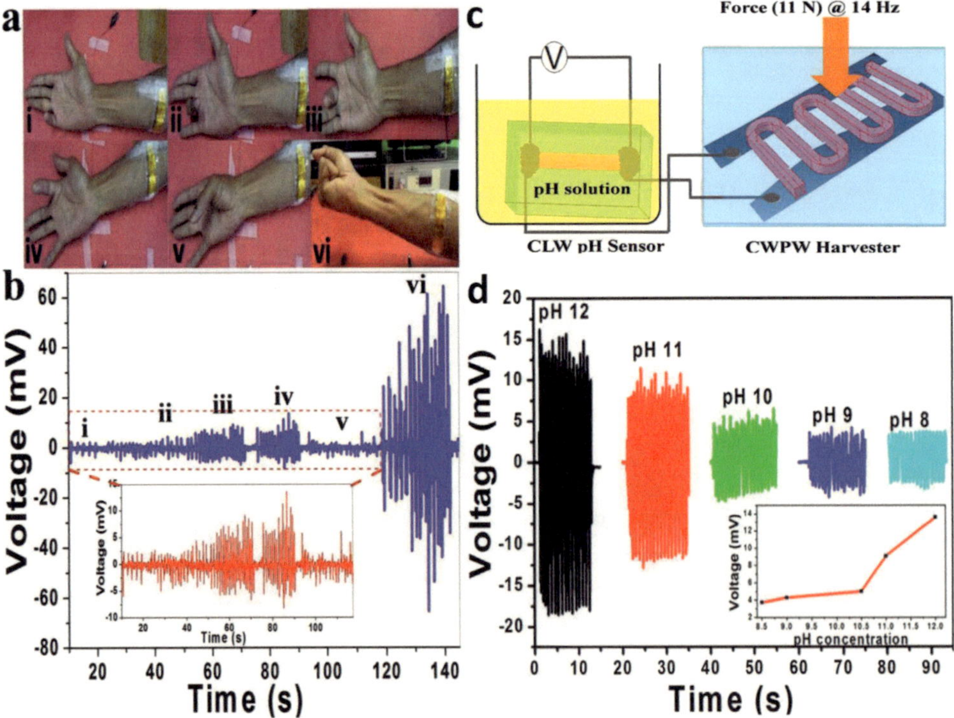

Figure 6. (a) Demonstration of the SFS to measure individual human finger flexion/movements and the strip-type HSBbased PNG device mounting on the forearm of the human hand. (b) The electrical response of the SFS concerning the finger motions and the inset shows the magnified electric response of SFS. (c) Schematic represents the demonstration of the SPS using the parallel connection between the HWPW-based PNG device and the HLW pH sensor. (d) The electrical response of the SPS under various alkaline solutions ranges from a value of 12 to 8. The figures are reproduced with the permission from Refs. [17, 18]. Copyright of Elsevier.

http://dx.doi.org/10.5772/intechopen.74770

In the case of SFS, the finger motion generates the small amount of mechanical force through the tendon-muscle unit of the forearm, which perpendicularly acts on the SFS results in the change in electrical response concerning the individual finger. Here, the location of the SFS on the forearm, length, and sensitivity of the SFS will determine the perfect classification and measurement of the individual fingers (thumb, index, middle, ring, and pinky) as shown in **Figure 6b**. Whereas in the case of the SPS, HWPW-based device works as a sustainable independent power source to drive the HLW pH sensor as shown in **Figure 6c**. The output voltage of the HLW pH sensor decreases from 33 to 7.3 mV, when the pH buffer solution changed from 12 to 8. It indicates that the change in resistance of the pH sensor decreases gradually concerning the decrease in the pH value of buffer solution as shown in **Figure 6d**.

To measure the nonlinear surface motions of any object or biological body requires the highly adaptable, flexible, and having a small device area. For this, efficient, adaptable HPHS are fabricated by the invention of the cost-effective GT. Here, the hybrid strips are possible to make any shape depending on the type of the groove used, proper hybrid (polymer + inorganic $0.3Ba_{0.7}Ca_{0.3}TiO_3$–$0.7BaSn_{0.12}Ti_{0.88}O_3$ (BCST) NPs) solution. The synthesis process, crystalline phase information, ferroelectric hysteresis loop characterization, and the surface morphological studies of BCST NPS were given in our previous report [16]. **Figure 7a** shows the schematic diagram of the detailed process of the GT for the HPHS explained in the experimental section. The crystalline phase of the HPHS (PDMS/BCST NPs), pure PDMS, and BTO NPs was investigated by the Raman spectra as shown in **Figure 7b**. Briefly describing, the peak

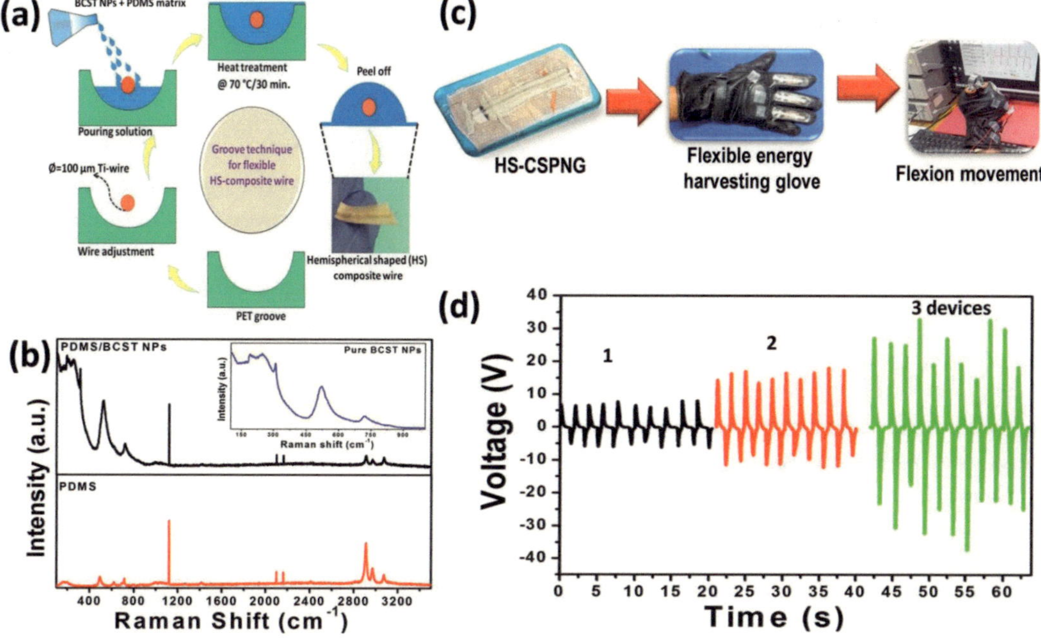

Figure 7. (a) Cost-effective grove technique (GT) for preparing the HPHS and its optical photograph. (b) Raman spectra of the pure PDMS, HPHS and the inset shows the high crystalline BTO NPs. (c) Optical photograph of the as-developed hemispherical composite strip-type PNG (HS-CSPNG) and the development of flexible energy-harvesting glove (using three devices) and its flexion movement during the electrical measurement. (d) Output voltage response of flexible energy harvesting glove upon the flexion/extension movements of human hand fingers. The figures are reproduced with the permission from Ref. [16]. Copyright of Royal Society of Chemistry.

between 300 and 310 cm^{-1} in HPHS and BCST NPs was confirmed by the existence of the higher tetragonal phase content. Whereas the peaks in PDMS spectra confirm the existence of the CH_3 (symmetric, asymmetric stretching's), Si—C (symmetric), Si—CH_3 (symmetric), and Si—O—Si (symmetric) vibration bands. Further, as-developed HPHS are used to fabricate the HS-CSPNG devices, and flexible energy-harvesting glove and the corresponding digital photographs are shown in **Figure 7c**. The flexible energy-harvesting glove was tested to harness the waste biomechanical energy, that is, harnessing the flexion/extension movements of the finger converted into a useful electrical signal. The single HS-CSPNG device generates an approximately peak-to-peak voltage of ≈10 V, a series connection of two HS-CSPNGs and three HS-CSPNGs generates (≤30 V) and (≤60 V) upon flexion/extension movements of the finger (**Figure 7d**). It demonstrates that the as-developed HPHSs and the HS-CSPNGs are highly potential candidates to harness the waste mechanical energy, having a reduced device area and low cost. Further, these HPHS are used to develop the self-powered muscle-monitoring system to measure the nonlinear surface motions of the human body parts at a time without using any additional storage battery, circuit components (reference). Other than these, as-developed PVDF/AC hybrid film was tested to harness the mechanical energy and also used to develop the self-powered acceleration sensor to measure the various accelerations of the linear motor shaft load. The detailed fabrication protocol, characterization analysis, and electrical responses were given in our previously published report [20]. The described basic approaches for piezoelectric nanogenerators and self-powered sensors in the present work can pave the way to develop the next-generation flexible/portable smart electronics/sensors.

4. Conclusions

In summary, the lead-free piezoelectric $BaTiO_3$ NPs, NCs, and BCST NPs along with the polymer support can be possible to generate the innovative HPSs to improve the instantaneous power density of the PNG. The flexible HPSs performance depends on the weight ratio of nanoparticles in a polymer matrix, thickness of the film, external poling process, and the applied perpendicular mechanical force. The fabrication methods for HPSs are cost-effective, eco-friendly, simple, and possible for the large-scale production. The synergistic (piezoelectric-triboelectric) effect at the material design level or the device design level will enhance the generated power density of the device. Few times, as-fabricated PNGs have a dual functionality such as harnessing the mechanical energy (sustainable independent power source) and can also work as a self-powered sensor without using additional storage components. The results demonstrate that the HPSs and its nanogenerators are potential candidates to develop the self-powered sensors/systems to monitor the physical/optical/chemical stimuli and glow the low-power consumed LEDs.

Acknowledgements

This work was supported by the Jeju Sea Grant College Program 2017, funded by the Ministry of Oceans and Fisheries (MOF), and by the National Research Foundation of Korea (NRF), funded by the Korea Government Grant (2016R1A2B2013831).

Conflict of interest

There are no conflicts to declare.

Author details

Nagamalleswara Rao Alluri[1], Arunkumar Chanderashkear[2] and Sang-Jae Kim[2*]

*Address all correspondence to: kimsangj@jejunu.ac.kr

1 Mechanical Engineering, Faculty of Applied Energy System, Jeju National University, Jeju Island, South Korea

2 Department of Mechatronics, Nanomaterials and Systems Lab, Jeju National University, Jeju Island, South Korea

References

[1] International Energy Outlook 2017. Available from: https://www.eia.gov/outlooks/ieo/pdf/0484(2017).pdf [Accessed: 30-01-2018]

[2] Energy crisis, causes of the energy crisis and possible solutions of the energy crisis. Available from: https://www.conserve-energy-future.com/causes-and-solutions-to-the-global-energy-crisis.php [Accessed: 30-01-2018]

[3] Yuan-Ping L, Dejan V. Self-powered electronics for piezoelectric energy harvesting devices (chapter-14). In: Lallart M, editor. Small-Scale Energy Harvesting. London, United Kingdom: InTechopen Limited; 2012. ISBN: 978-953-51-0826-9. DOI: 10.5772/51211

[4] Wang ZL, Wu W. Nanotechnology enabled energy harvesting for self-powered micro/nanosystems. Angewandte Chemie, International Edition. 2012;**51**:2-24. DOI: 10.1002/anie.201201656

[5] Venkateswaran V, Nagamalleswara Rao A, Yuvasree P, Arunkumar C, Sang-Jae K. A flexible, planar energy harvesting device for scavenging roadside waste mechanical energy via the synergistic piezoelectric response of $K_{0.5}Na_{0.5}NbO_3$-$BaTiO_3$/PVDF composite films. Nanoscale. 2017;**9**:15122-15130. DOI: 10.1039/C7NR04115B

[6] Wang ZL, Song J. Piezoelectric nanogenerators based on zinc oxide nanowire arrays. Science. 2006;**312**:242-246. DOI: 10.1126/science.1124005

[7] Arunkumar C, Nagamalleswara Rao A, Saravanakumar B, Sophia S, Sang-Jae K. Human interactive triboelectric nanogenerator as a self-powered smart seat. ACS Applied Materials & Interfaces. 2016;**8**:9692-9699. DOI: 10.1021/acsami.6b00548

[8] Ya Y, Wenxi G, Ken CP, Guang Z, Yusheng Z, Yan Z, Youfan H, Long I, Wang ZL. Pyroelectric nanogenerators for harvesting thermal energy. Nanoscale. 2012;**12**:2833-2838. DOI: 10.1021/nl3003039

[9] Ya Y, Ken CP, Qingshen J, Jyh Ming W, Fang Z, Yusheng Z, Yue Z, Wang ZL. thermoelectric nanogenerators based on single Sb-doped ZnO micro/nanobelts. ACS Nano. 2012; **6**:6984-6989. DOI: 10.1021/nn302481p

[10] Joe B, Steve D. Piezoelectric nanogenerators—A review of nanostructured piezoelectric energy harvesters. Nano. 2015;**14**:15-29. DOI: 10.1016/j.nanoen.2014.11.059

[11] Qiang Z, Bojing S, Li Z, Wang ZL. Recent progress on piezoelectric and triboelectric energy harvesters in biomedical systems. Advancement of Science. 2017;**4**:1700029. DOI: 10.1002/advs.201700029

[12] Kwi-Il P, Chang KJ, Na Kyung K, Lee KJ. Stretchable piezoelectric nanocomposite generator. Nano Convergence. 2016;**3**:1-12. DOI: 10.1186/s40580-016-0072-z

[13] Nagamalleswara Rao A, Saravanakumar B, Sang-Jae K. Flexible, hybrid piezoelectric film ($BaTi_{(1-x)}Zr_xO_3$)/PVDF nanogenerator as a self-powered fluid velocity sensor. ACS Applied Materials & Interfaces. 2015;**7**:9831-9840. DOI: 10.1021/acsami.5b01760

[14] Saravanakumar B, Thiyagarajan K, Nagamalleswara Rao A, Shin SY, Taehyun K, Lin ZH, Sang-Jae K. Fabrication of an eco-friendly composite nanogenerator for self-powered photosensor applications. Carbon. 2015;**84**:56-65. DOI: 10.1016/j.carbon.2014.11.041

[15] Nagamalleswara Rao A, Sophia S, Arunkumar C, Saravanakumar B, Jeong JH, Sang-Jae K. Piezoelectric $BaTiO_3$/aginate spherical composite beads for energy harvesting and self-powered wearable flexion sensor. Composites Science and Technology. 2017;**142**:65-78. DOI: 10.1016/j.compscitech.2017.02.001

[16] Nagamalleswara Rao A, Venkateswaran V, Arunkumar C, Saravanakumar B, Sang-Jae K. Adaptable piezoelectric hemispherical composite strips using a scalable groove technique for a self-powered muscle monitoring system. Nanoscale. 2018;**10**:907-913. DOI: 10.1039/C7NR06674K

[17] Nagamalleswara Rao A, Sophia S, Arunkumar C, Saravanakumar B, Lee GM, Jeong JH, Sang-Jae K. Worm structure piezoelectric energy harvester using ionotropic gelation of barium titanate-calcium alginate composite. Energy. 2017;**118**:1146-1155. DOI: 10.1016/j.energy.2016.10.143

[18] Nagamalleswara Rao A, Sophia S, Arunkumar C, Saravanakumar B, Jeong JH, Sang-Jae K. Self-powered pH sensor using piezoelectric composite worm structures derived by ionotropic gelation approach. Sensors and Actuators B: Chemical. 2016;**237**:534-544. DOI: 10.1016/j.snb.2016.06.134

[19] Nagamalleswara Rao A, Arunkumar C, Venkateswaran V, Yuvasree P, Sophia S, Ji Hyun J, Sang-Jae K. Scavenging biomechanical energy using high-performance, flexible $BaTiO_3$ nanocube/PDMS composite films. ACS Sustainable Chemistry & Engineering. 2017;**5**:4730-4738. DOI: 10.1021/acssuschemeng.7b00117

[20] Nagamalleswara Rao A, Arunkumar C, Ji Hyun J, Sang-Jae K. Enhanced electroactive b-phase of the sonication-process-derived PVDF-activated carbon composite film for efficient energy conversion and a battery-free acceleration sensor. Journal of Materials Chemistry C. 2017;**5**:4833-4844. DOI: 10.1039/c7tc00568g

Chapter 3

Wearable Human Motion and Heat Energy Harvesting System with Power Management

Juris Blums, Ilgvars Gornevs, Galina Terlecka,
Vilnis Jurkans and Ausma Vilumsone

Additional information is available at the end of the chapter

http://dx.doi.org/10.5772/intechopen.74417

Abstract

A combined human motion and heat energy harvesting system are under investigation. Main parts of the developed human motion energy harvester are flat, spiral-shaped inductors. Voltage pulses in such flat inductors can be induced during the motion of a permanent magnet along its surface. Due to the flat structure, inductors can be completely integrated into the parts of the clothes, and it is not necessary to allocate extra place for movement of the magnet as in usual electromagnetic harvesters. Prototypes of the clothing with integrated proposed electromagnetic human motion energy harvester are created and tested. Voltage of generated impulses is shown to be high enough to be effectively rectified with commercially available diodes and ready to be stored; however, efficiency depends on properties of controlling circuit. In order to increase the sustainability of the energy source and its stability, an option for combining a motion energy harvester with a human body heat energy harvester is also considered. Thermoelectric generator that harvests electricity from waste heat of human body is presented, and generated voltage and power are compared at different activity levels and ambient temperatures. Power generated with thermoelectric generator located on lower leg reached up to 35 mW with peak voltages reaching 2 V at certain conditions. A possible power management set-up and its efficiency are discussed.

Keywords: human motion energy harvesting, smart apparel, flat inductors, thermoelectric generators, microwatt power management

1. Introduction

Electronic devices are becoming smaller, lighter, and consume less power. These tendencies cause search for energy sources that will produce electricity using energy that is available in

our surrounding environment (light, motion, heat, etc.) [1]. With such energy source, electronic device would become independent and there would be no need to periodically access the battery to change or charge it. It would also allow designing embedded systems, which will be protected from harmful environmental influence. Wearable electronic systems integrated in clothing elements are one of the device groups that need to be protected [1, 2]. Aims and tasks for such systems are security and monitoring of physiological parameters [3, 4], determination of location [5], and communication [6, 7]. At the moment, practically all such systems are based on integration of various sensors in clothing to gather information about surroundings and send it to remote receivers for data storage and processing, as well as to inform person about the changes in the environment or any harm or danger [3]. Powering of such systems is possible by using electrochemical batteries [8, 9], solar cells [10, 11], thermoelectrical elements [12], converters of mechanical energy (piezoelectric [1, 13], electrostatic [1], and electrodynamic [1, 14] energy harvesters). Cost of change or charging as well as necessity to access electrochemical elements are considered as major disadvantages. Mentioned alternative energy sources use various physical phenomena and can be combined to work together, optimizing work of wireless electronic devices. Mechanical energy harvesters are regarded as most perspective and universal energy source for wearable electronics according to various scientists [1, 15]. Multiple publications offer such harvesters for footwear [16], add-on for bags or purses [14], or as accessories [17]. In almost all cases, the construction of these harvesters considers either human body part direct force [16] or periodic motion [14] conversion into electric energy. Most of prototypes are three-dimensional devices [14, 18], which makes it impossible to implement them into clothing without changing its functionality and appearance.

Authors and other researchers have previously shown that it is possible to use flat spiral-shaped coils [19, 20] as inductors in electrodynamic harvesters and elements of electronic circuits. Another advantage of electromagnetic harvesters is their ability to work without direct contact between harvester elements, as it is necessary, for example, with piezoelectric harvesters; therefore, it is possible to implement them without changing shape, elasticity, and other mechanical properties of clothing [19].

It is possible to use different technological methods when designing wearable electronic elements [21, 22], which allows to make flexible and stretchable connections for components. One of the technologies that can be used for electronic connections is embroidery, which can be considered as one of the most perspective ways to make conductive structures in textiles and on the surface of clothing elements. It is possible to make structures (connections, resistive, and inductive elements) that preserves properties of fabric and only minimally changes clothing shape. Recently embroidery was suggested as a method to make inductive elements for resonant circuits [21].

Another perspective wearable energy harvesting method is the use of thermoelectric generators, which use waste heat as energy source. Due to rigidity, these harvesters are mainly used for such wearables as wristwatches [23, 24]. However, commercially available thermoelectric elements are available in various sizes; therefore, they can be used in clothing as well. The main advantage of these harvesters is that they provide direct current (DC), which can be generated in both high and low activities of a person who wears it.

Work [19] presents possibility to design clothing and human body part movement energy transformer, which generated electrical energy using motion of permanent magnet in parallel with flat spiral inductors. Design, properties, and performance of such generators as well as possibilities to implement these generators into various clothing elements are presented and discussed in the following parts of this chapter.

2. Analysis of electromotive force and energy generated in inductive elements in homogenous magnetic field

Proposed human motion energy harvester working principle is based on electromagnetic induction phenomenon, when electromotive force (EMF) ε and current I in closed loop is induced by changing flux of the magnetic field Φ_M.

Assuming wire inductance is negligible, self-inductive electromotive force (EMF) will not be generated. Induced EMF can be calculated according to Faraday's law: electromotive force is proportional to the rate of change of magnetic flux Φ_M:

$$\mathcal{E} = \frac{d\,\Phi_M}{dt} = \frac{d(BScos\alpha)}{dt} \tag{1}$$

Supposing that magnetic field is homogenous ($B = const$) and perpendicular to coil plane ($\alpha = 0$), rate of change of magnetic flux will be proportional to crossed area changing rate:

$$\mathcal{E} = B\frac{dS}{dt} \tag{2}$$

where ε is the electromotive force [V], B is the magnetic induction [T], S is the magnetic field crossed conductor loop area [m^2], and t is the time [s].

If load resistance is much higher than resistance of winding, it can be assumed that there is no significant current flow, so generated voltage is equal to EMF. This is not optimal load condition and can be only used to characterize output voltage for ideal case. In reality, smaller load resistance leads to current increase, but it also decreases output voltage because every real generator has internal resistance (dependent on used materials, construction, etc.), which dissipates part of generated energy on itself. For this reason, it is possible to find the optimal condition for real generator, where load receives the highest energy. It happens when load resistance value is equal to internal resistance of generator. In such case, voltage on load and internal resistance is equal to half of EMF; they both share the same dissipated energy. Total energy available on the resistive load according to the Joule-Lenz law is:

$$W = \int Pdt = \int \frac{U^2}{R}dt \tag{3}$$

where U is the voltage on resistive load [V]; R is the resistance of the load [Ω]; $P = U^2/R$ is the instantaneous power [W].

By using analytical expression for spiral shaped, for example, Archimedean spiral, induction element, or element group, it is possible to apply Eq. (2) for specific instance. Generated EMF will depend on the magnet velocity v and shape of the inductor, which is crossed by magnetic field:

Shape	Induced EMF	Generated energy	Energy (rationed) according to theory	Energy (rationed) in experiments
Square	$\mathcal{E} = Bav$	$W = \frac{B^2 a^3 v}{R_1}$	1	1
Rhombic	$\mathcal{E} = B v^2 2t \cdot tg(\alpha)$	$W = 0.94 \frac{B^2 a^3 v}{R_2}$	0.94	0.93
Circular	$\mathcal{E} = 2rvB\sqrt{1 - \left(1 - \frac{vt}{r}\right)^2}$	$W = 0.66 \frac{B^2 a^3 v}{R_3}$	0.85	0.79

Table 1. Results of theoretical calculations and experimental results for inductors with different geometrical shapes [25].

$$\mathcal{E} \sim \frac{dS}{dt} = f(v, \text{shape})$$

Ref. [25] shows analysis of power and energy for different shapes of flat inductor (square, rhombic, and circular) and theory is backed up with experimental data. Square and rhombic edge length *a* were considered equal to the diameter of round wire *2r* to obtain data in **Table 1**.

Theoretical analysis of different shapes shows that the highest energy should be obtained from square-shaped coil (magnet size is assumed to be equal to coil size). Experimental data for various sizes and shapes of inductors proved theoretical predictions. Relative energy and equations are shown in **Table 1**.

3. Experimental: motion energy harvester

3.1. Flat inductors

Flat inductors can be manufactured in various ways, many of which are related to types of clothing manufacturing technology. Here are some ways to produce flexible inductive coils for proposed harvesting:

1. Printed circuit board (PCB) technology on elastic substrate. This method can be used to create inductors of any shape with high resolution and also combine some parts of power management system on a flexible substrate (**Figure 1**). Since the final product is very thin, integration into apparel is easy and done without significant alterations, but the same reason creates significant coil resistance as copper tracks are long and usually thinner than 100 μm to keep the whole structure flexible. Even though longer conductor track produces higher voltage, larger internal resistance (over a hundred of ohms) reduces power output significantly.

2. Various shapes and sizes can be created by using embroidering by conductive yarns. The test structure (**Figure 2**) has the upper (needle) polyester thread and the lower (bobbin) conductive thread, which is a silver-coated polyamide thread Elitex 110 dtex/f34x2. The length of the tested square-shaped coil is 25 mm, and it consists of 19 windings. The main disadvantage as compared to PCB technology is the resolution as winding density is much lower which leads to lower generated voltages.

Figure 1. Flat spiral coil, made of 35 μm thick copper tracks and polyurethane substrate (created in *Fraunhofer Institute for Reliability and Microintegration IZM, Berlin, Germany*); length of an edge is 25 mm.

3. A thin copper wire can also be sewed on a fabric, for example, with zigzag stitch technology by using embroidering machine like SZK JCL 0100-585. Wires are much more fragile than threads, but this technology puts noticeably lower stress on the wire during production as the wire is supplied by a dedicated device and fixed to a fabric with zigzag stitch. The

Figure 2. Embroidered flat square-shaped (length of an edge is 25 mm) inductor.

lower internal resistance is an additional advantage of such a method in comparison with the abovementioned. **Figure 3** shows Archimedean spiral with 25 mm diameter, 25 windings with a 0.2 mm copper wire. Stitching step must be adjusted carefully to protect the wire and its insulation layer from accidental damage by needle.

4. Coils can be made as a design element by machine production or handcrafting. Article [26] shows multiple examples of creation of inductors by crocheting (**Figure 4**) and knitting.

5. Manually made coils in Archimedean spiral shape [19] were the ones used in further discussed prototypes. All of them were 25 mm in diameter and winded using 0.22 mm copper wire, except coils mentioned in **Table 2**.

3.2. Number of flat inductors in the generator and its testing

To achieve maximal performance of the generator and acquire the highest amount of energy, it is necessary to place several inductors into elements of clothing in a way that magnet could induce electromotive force in several inductors sequentially. It is possible in such places where one part of clothing moves as fast and close to other parallel part as possible. Works [25, 27] analyze dependence of voltage impulses with regard to their placement to find maximal count of inductors that are allowed to be placed in clothing in path of movement of one magnet. By placing four inductors on magnet trajectory and analyzing relative increase of momentary power in respect to inductor placement, it was shown that it is reasonable to position two or three inductors so that their generated voltage impulses add with each other.

Figure 3. Copper wire sewed on a textile substrate (created in RWTH Aachen University – Institut für Textiltechnik, Aachen).

Figure 4. Crocheted flat inductor [26].

Coil type	Internal resistance, Ω	Mean energy, μJ	Mean crossing time, s	Mean power, μW	Notes
Embroidered coil	8.3	2.21	0.075	29.84	Diameter 25 mm with 19 windings
Sewn coil with zigzag stitch technology	0.7	6.62	0.072	91.25	Diameter 25 mm, copper wire diameter 0.2 mm, 25 windings
Manually made coil	2.2	113.82	0.130	875.54	Diameter 30 mm, copper wire diameter 0.22 mm, 60 windings
PCB coil	116	4.31	0.146	29.62	Diameter 30 mm, 80 windings

Table 2. Experimentally defined performance of inductors created by different techniques.

Mechanical manipulator was used to compare different inductors, generated electromotive force, and overall energy. It provided controlled and repeatable conditions for tests of various inductors. Mechanical manipulator moved square-type permanent magnet ($Nd_2Fe_{14}B$, 0.2 T, 20 × 20 × 2 mm) along spiral flat inductors with controlled velocity. Voltage impulses were registered for each tested inductor on resistive load equal to inductors' internal resistance. Generated power and energy were defined using Eq. (3).

Table 2 shows experimental data of flat inductors that are made in various ways. It can be seen that embroidered coils have less windings compared to manually made coils as well as relatively high internal resistance. This is the reason why energy and power are low in comparison with others. The sewn coil energy and power values are smaller than those of the manually created coil, which is explained by the fact that such coils have two times lower number of windings, which cannot be increased by using manufacturing techniques.

Power from PCB technology coil falls behind manually winded coil as tracks are very thin, introducing significantly larger internal resistance, even though generated voltage is larger (up to 0.2 V for a single coil) due to larger number of windings. Manually made coils produce about 30 times higher power, but voltage reaches only 0.12 V on the load resistance.

3.3. Prototypes of clothes with energy harvesters

3.3.1. General considerations

Previously done tests with automatic magnet movement with constant parameters provide data for different inductor comparison, but the actual environment for proposed harvesters is far from constant. Multiple clothing prototypes were created for real-life results, two of them are described in the following sections. Both were tested on a person with different walking speeds.

Since the requirements for highest voltage and power are relative to the highest speed of the magnet's motion and as small as possible distance between the magnet and the inductor, the most suitable place is a sleeve and side of an outerwear. These positions are shown in **Figure 5**.

3.3.2. Prototype 1

Energy harvester with planar structure integrated into the men's insulated outerwear and generator is located at the anterior superior iliac spine level (**Figure 6**). The location of the inductive elements in the prototype 1 is marked in **Figure 5** by *1* and the location of the magnet is indicated by 2. Insulated in outerwear, integrated electromagnetic human motion energy harvester contains two parts:

- The inductive element consists of three groups of spiral-shaped coil with identical direction of winding turns, which are connected in series (coil groups have 1 cm space between them and generated voltages gave the same polarity). Each coil group consists of five layers, placed one onto another with insulating layer in between, coils have 2.5 cm diameter and 50 windings.
- The second generator part is lightweight, small, and strong neodymium (Nd) magnet with double magnetic field structure and induction 0.26 T. An arc-shaped magnet is 1/6 from concentric circles with r_1 = 2.5 cm and r_2 = 4.0 cm.

The volume of the generator (coils + magnet) is approximately 4.63 cm^3. The total mass of the integrated elements is about 50 g.

The generator was tested by a wearer during the process of walking at fixed speeds (**Figure 6**). The typical shape of the pulses of generated voltage during the periodic motion of hands along the wearer's body are shown in **Figure 7**.

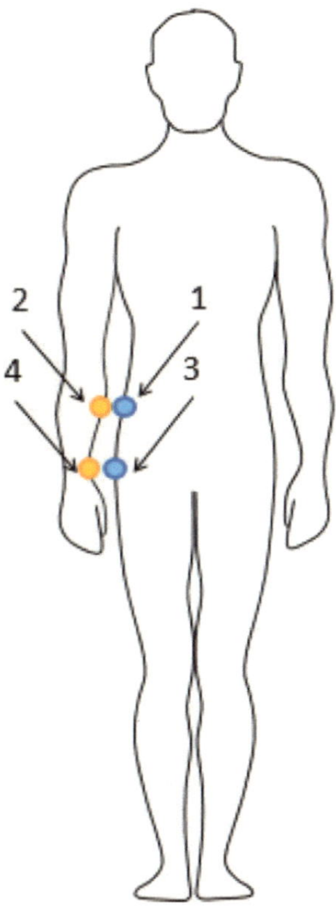

Figure 5. Positions of parts of electromagnetic harvester. 1 and 3 – Positions for flat inductors, 2 and 4 – For the permanent magnet.

The voltage generated during one full motion cycle of the magnet along the coil consists of two pulses. The full motion cycle—the period of double step—is formed for each arm from both forward and reverse movements. Some asymmetry of pulses connected with the trajectory of sleeve movement was observed. It was observed that during the movement of sleeve forward, trajectory of magnet is maximally close to the coils and voltage generated in pulse is higher (left pulse in **Figure 7**), while during movement of sleeve backwards, magnet is further away from them and generated voltage is lower (right pulse in **Figure 7**). Impulses are not symmetrical due to natural hand movement and clothing deformation.

The numerical data of harvester-relevant characteristics for the full walk cycle are shown in **Table 3**. Maximal mean power 0.50 ± 0.10 mW is observed at a walking speed of 6 km/h.

3.3.3. Prototype 2

In a men's jacket (prototype 2), energy harvester elements are integrated at the carpal joints level, maintaining symmetry (**Figure 5**, positions 3 and 4). The energy harvester's parts in the prototype 2 are the following:

- inductor array: three sets of coils are placed near to each other and connected in a series in a way that generated voltage for each set has opposite polarity;

Figure 6. Testing prototype 1 on treadmill.

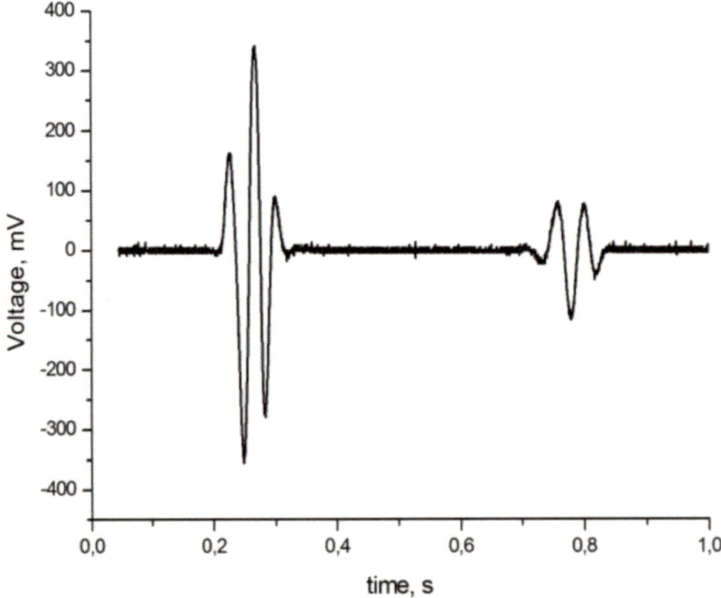

Figure 7. Prototype generated voltage pulses at walking speed of 6 km/h.

Walking speed, km/h	Mean energy, mJ	Time of 1 walking cycle, s	Mean power, mW	Mean power density, mW/cm³
3.0	0.23	1.27	0.18	0.04
4.5	0.29	1.10	0.26	0.06
6.0	0.49	0.98	0.50	0.11

Table 3. Prototype 1 in use—Performance.

- magnetic field source: two round magnets with 20 mm diameter and 0.2 T magnetic induction.

Volume of the generator (coils + magnet) is approximately 4.95 cm^3. The total mass of integrated elements is 30 g. The voltage pulse generated is shown in **Figure 8** (**Table 4**).

3.4. Power management

As generated pulses have low amplitude, even such a simple and important task as rectification brings in some challenge. To verify usability of prototype 2, it was tested against typical full-bridge rectifier with DFLS120L Schottky diodes and resistive load (**Figure 9**), in which the values were changed (10, 30, 50, 70, 100, and 200 Ω). Voltage was measured before and after the rectifier (**Figure 8**) to get information about energy lost on the rectification circuit.

The most effective load for any generator is the one with value equal to internal resistance of the generator itself. This is because the energy on the load is the highest at these circumstances.

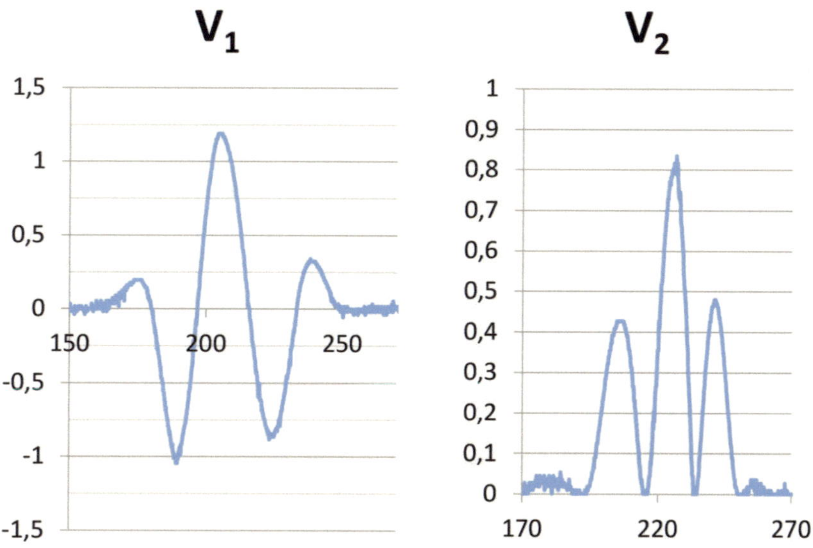

Figure 8. Voltage before rectifier (V_1) and on 70 Ω load (V_2); x axis — Time (ms), y axis — Voltage (V).

Walking speed, km/h	Mean energy, mJ	Time of 1 walking cycle, s	Mean power, mW	Mean power density, mW/cm³
3.0	0.24	1.35	0.18	0.036
4.5	0.30	1.28	0.23	0.046
6.0	0.49	1.11	0.44	0.088

Table 4. Prototype 2 in use—Performance.

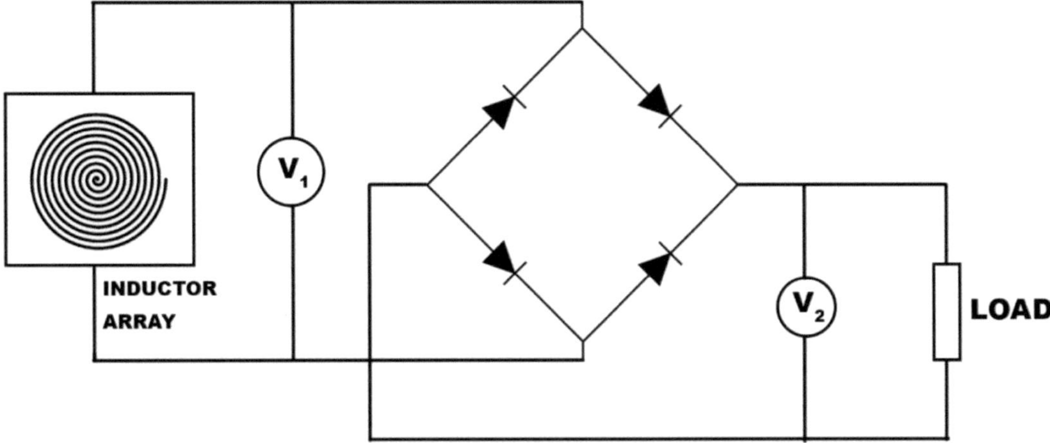

Figure 9. Measurement schematic. V_1 is measured voltage before rectifier (input voltage), V_2 is output voltage on the load.

As we have additional interface (bridge rectifier) between the load and the generator and it is changing its parameters depending on the generated voltage, multiple load values were tested. Internal resistance of the prototype 2 was 17 Ω, voltage measurements were done on the generator and load at the same time.

To evaluate generator effectiveness in prolonged period, 40 consecutive impulses were recorded during the normal movement of a test subject. True root mean square (RMS) values for every two impulses (1 s) of input and output voltages were calculated, resulting an average value and standard deviation of distinct measurements. By known load value, RMS current was calculated, that was used for generated power estimation as the current is the same for input and output. Each true RMS calculation was done for time period of a full second, thus numerically power from this RMS value is equal to energy. Results are provided in **Table 5**.

Standard deviation for voltage RMS measurements is small enough to consider average values for current measurements being true. Average power and energy is the product of RMS voltage and current on input and output. Efficiency increases with load resistivity because higher load value leads to smaller current in circuit, which decreases voltage drop on rectification elements, thus reducing wasted energy. On the other hand, it also decreases output power; accordingly, a trade-off must be chosen between efficiency and high enough output voltage and current as real-life load like sensors, batteries, and other systems will depend on that.

	10 Ω	30 Ω	50 Ω	70 Ω	100 Ω	200 Ω
RMS on load (mV)	57.77 ± 1.14	98.65 ± 2.28	106.99 ± 2.24	133.59 ± 2.02	157.54 ± 3.43	168.83 ± 1.83
RMS on input (mV)	168.66 ± 1.64	204.01 ± 2.23	206.43 ± 2.71	227.87 ± 2.47	242.50 ± 4.54	247.7 ± 2.85
RMS current (mA)	5.78	3.29	2.14	1.91	1.58	0.84
Output power/energy for an impulse (μW, μJ)	166.84	162.19	114.47	127.47	124.09	71.26
Input power/energy for an impulse (μW, μJ)	487.13	335.43	220.86	217.44	191.02	104.55
Efficiency (%)	34	48	52	59	65	68

Table 5. Prototype 2 rectification experimental data. RMS calculated for time period of 1 s (two impulses).

4. Combination of the motion energy harvester with the human body heat energy harvester

Taking into account the fact that electromagnetic harvester will generate electricity only when a person that wears it is moving, its energy might be insufficient to continuously supply power to wearable electronic systems. To increase overall energy supply, it can be useful to combine energy harvesters that rely on different kinds of waste energy from human body. One of the kind might be thermal energy that can be converted to electricity using thermoelectric elements, for example, Peltier elements. Various studies [28–30] have shown that use of thermoelectrical generators may be a successful solution for harvesting energy for wearable electronic devices.

To evaluate possible gain of combining electromagnetic and thermoelectric generators, it was decided to build prototype using thermoelectric elements.

Thermoelectric harvester prototype with 5 Peltier elements was made to convert waste heat of human body into electrical energy (**Figure 10**). All five elements were connected in series to achieve higher generated voltage. System is mechanically connected using elastic strings to apply it to naked skin on arm or leg. Total surface area taken by Peltier elements is 4500 mm^2, which is optimal to cover forearm or lower leg.

Measurements were performed in two different environments:

1. Indoors—activities inside the building at 20°C ambient temperature, low air flow
2. Outdoors—activities outside at 5°C ambient temperature, wind speed approximately 3 m/s.

Generator was tested both covered with clothing and uncovered, during three levels of activity: "no activity"—the person is at rest, "low activity"—walking, and "high activity"—running.

Table 6 represents both open circuit voltage (U_{OC}) and power (P) at balanced load:

Measurements reveal that even in quite difficult for the heat exchange conditions (indoors, covered, no activity) with generator based on 5 Peltier elements, it is possible to generate voltage up to 55 mV, which is enough to convert it to higher voltages with Step-Up converter, for example, LTC3108 manufactured by Linear Technology. This means it is possible to get continuous

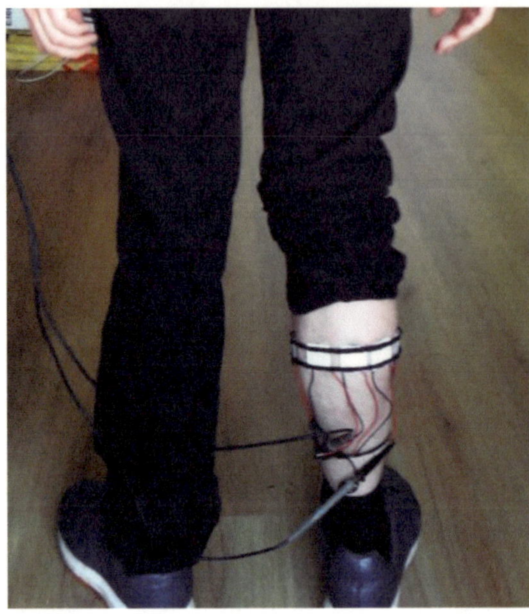

Figure 10. Thermoelectric generator on lower leg.

	Indoors				Outdoors, cold weather			
	Covered		Uncovered		Covered		Uncovered	
Type of activity	**U_{OC}, V**	**P, mW**	**U_{OC}, V**	**P, mW**	**U_{OC}, V**	**P, mW**	**U_{OC}, V**	**P, mW**
No activity	0.065	0.05	0.09	0.1	0.145	0.28	0.34	1.5
Low activity	0.11	0.21	0.172	0.41	0.195	0.51	0.43	2.45
High activity	0.14	0.42	0.26	0.93	0.234	0.74	0.46	2.97

Table 6. Performance of the thermoelectric generator.

50 μW power from the generator all the time while it is on a wearer. This power is enough to power devices such as wearable wireless sensors for physiological signal monitoring [31].

If conditions are good for generator (outdoors, uncovered, high activity), it is possible to obtain up to 3 mW power, which could be enough for activities such as wireless data transmission. This is contributed by both higher temperature difference between skin and air as well as good airflow, resulting in higher heat transport through Peltier elements.

Most probable results that should be expected for person with moderate activity during the day would be 0.1–0.2 mW on average, which is typical power consumption for high-performance general purpose microcontrollers [32].

Power generated by both electromagnetic and thermoelectric harvesters are similar, therefore both can provide comparable energy to be stored to single capacitor or battery. Taking into account the discontinuous nature of each generator, storing energy in the same element might result in less interruption in the supply of energy.

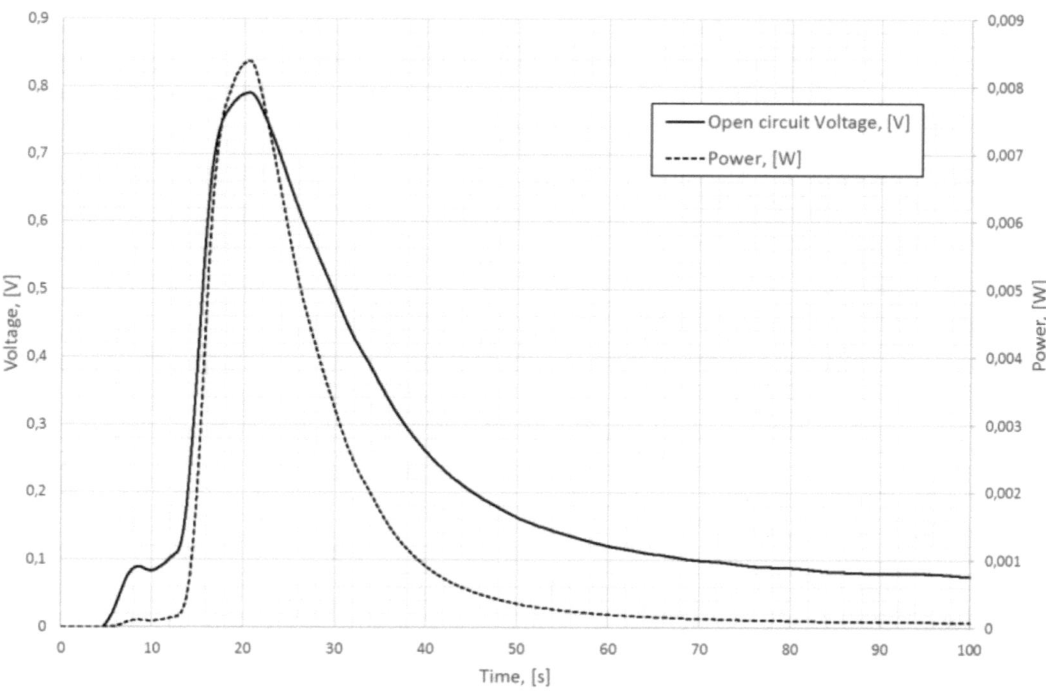

Figure 11. Open circuit voltage and power generated by thermoelectric generators during putting on.

An interesting moment of such thermoelectric generator that could be quite useful is the moment of putting it on—when generator, that has ambient room temperature, is exposed to warm skin, at least 10°C difference between sides of Peltier elements is produced. The generated voltage and the power developed by Peltier elements at this time are shown in **Figure 11**.

As we can see, peak power and voltage can be observed for about 10 seconds. Amount of generated voltage and power depends on ambient temperature, skin temperature, speed of putting the generator on, and several other factors. Power generated during this process fluctuates between 3 and 35 mW and voltages between 0.5 and 2 V. This can be quite useful energy, because start-up process of voltage converters that do not use external power source will be very slow if converter is being operated with lowest input voltages. For example, ultralow voltage step-up converter and power manager LTC3108 should be able to charge its output capacitor about 10 times faster if input voltage is 500 mV instead of 100 mV. It means that self-powered wearable electronic device will gather energy to start work much faster when using this advantage.

5. Discussion

Produced prototypes proved possibility to integrate human motion energy harvester into apparel without compromising its usability and characteristics. According to [33], this was the main problem of published energy harvesters developed for human motion energy harvesting.

As prototype 1 had thicker construction because of the materials and thermal insulation layer, harvester showed more stable results and higher generated energy than prototype 2. Thicker and more robust construction provides less free movement of individual garment parts, creating more distinct trajectory for harvester elements.

Considering necessity for integration into apparel, proposed flat inductor system has multiple advantages over other harvesters: flexibility, variable size, and harvester elements are located in different places and direct contact is not necessary, energy is harvested from human natural movements, and so on. Proposed harvester does not rely on inertia and can be used in broad movement frequency diapason with unchanged efficiency.

Since harvester parts are not meant to be enclosed in any rigid housing, there is no need to allocate any space for magnet movement like it is in traditional cylindrical coil construction [14], flat coils with magnetic suspension [34], or with the magnet moving through the coil [20]. Power density is considerably higher than for other electromagnetic harvesters summarized in [1], but most of them rely on inertia and thus have resonant frequency, which is high (e.g., $P/V = 2{,}2\ mW/cm^3$ at $f = 320$ Hz).

As human movement nature is individual and changes on many occasions, the same harvester can show different results for another clothing, user, or simply during time. Systems relying on inertial movement, for instance, MEMS, have much more stable performance. Prototype 1 shows 20% deviation in results, while prototype 2 up to 30%, invariance increases with faster walking pace. Generated energy is dependent on various factors, many of them are uncontrollable, for instance, specific walking features, pace, body load, and so on.

Comparing with other published electromagnetic harvesters, proposed harvester can generate high enough voltage to be rectified with off-the-shelf diodes with efficiency above 50%. On an average, peak output voltage is below 1 V, but such values can also be further used for low-power management systems, for example, voltage boost converter LTC3105 from Linear Technology starts up from 0.25 V (then works from 0.225 V) and can be programmed to load the harvester in a way to get the most effective power output. Such converter will start up on rectified pulses above 250 mV as they have relatively low frequency, it may give enough power at appropriate output voltage like 3.3 V for direct usage, but the most perspective objective is further accumulation in a battery or a capacitor. Some short activities like body or environmental parameter measurements might be powered directly, but it is unlikely due to unavoidable power losses in all power management steps.

Thermoelectric elements have several advantages when compared to electromagnetic motion harvesters: it generates electricity even when the person is not moving; relatively high voltages (up to several Volts) can be generated when system is applied to a person; generated electricity is direct-current (DC) type, therefore no rectifying is needed. These advantages make thermoelectric generators very beneficial addition to electromagnetic motion harvesters because (1) system that is powered by harvesters can be powered continuously, because thermoelectric generators will supply power, when it is not generated by movement; (2) startup, which is most difficult moment for every low-power harvesting system, can be done much faster and easier because higher energy generated by thermoelectric generators during putting on the wearable; and (3) generated DC voltage can be used as a bias voltage or additional power source to improve efficiency of rectifier used for electromagnetic generator.

6. Conclusion

The combination of human motion and heat energy harvesting systems is under investigation. The electromagnetic human motion energy harvester which consists only of flat elements and can be completely integrated into the apparel is under discussion. Main parts of the developed human motion energy harvester are flat, spiral-shaped inductors. Voltage pulses in such flat inductors are induced during the motion of a permanent magnet along its surface. By not using common method of magnet moving inside the coil but applying the proposed method with magnet and coil moving in parallel planes, new possibilities are created because such kind of movements precisely coincide with the relative motion of one part of apparel along the other. It is not necessary to keep hollow place for the movement of the magnet as in usual electromagnetic harvesters, because the parts of harvester are placed in separate elements of apparel. Due to the above-mentioned properties, the mean power density of developed harvester exceeds the values given for three-dimensional harvesters with traditionally used cylindrical coil. The prototypes of the clothing with integrated electromagnetic human motion energy harvester with flat inductors were tested. The power developed by the harvester during the walking with a speed of 6 km/h was up to 0.5 mW, which is suitable for powering wireless sensors. The voltage generated by the proposed harvester is shown as ready to be rectified with off-the-shelf diodes for further power management steps.

Thermoelectric harvester is proposed as addition to electromagnetic harvester. Combined system benefits not simply from extra power but from possibilities to differentiate power flow, for example, less powerful harvester can be used as independent source for control or power management system for the main harvester. As thermoelectric harvester does not fully depend on human movement, it can provide great support for motionless periods, creating more complete power harvester system.

Author details

Juris Blums[1]*, Ilgvars Gornevs[3], Galina Terlecka[2], Vilnis Jurkans[3] and Ausma Vilumsone[2]

*Address all correspondence to: blum@latnet.lv

1 Technical Physics Institute, Riga Technical University, Riga, Latvia

2 Institute of Design Technologies, Riga Technical University, Riga, Latvia

3 Institute of Radioelectronics, Riga Technical University, Riga, Latvia

References

[1] Mitcheson PD, Yeatman EM, Kondala RG, Holmes AS, Green TS. Energy harvesting from human and machine motion for wireless electronic devices. Proceedings of the IEEE. 2008;**96**(9):1457-1486

[2] Pulin G, Sarraute E, Costa F. Generation of electrical energy for portable devices. Comparative study of an electromagnetic and a piezoelectric system. Sensors and Actuators. A: Physical. 2004;**116**:461-471. DOI: 10.1016/j.sna.2004.05.013

[3] Bauer P, Sichitiu M, Istepanian R, Premaratne K. The mobile patient: Wireless distributed sensor networks for patient monitoring and care. In: IEEE EMBS International Conference on Information Technology Applications in Biomedicine, Arlington, VA, 2000. p. 17-21

[4] Coosemans J, Hermans B, Puers R. Integrating wireless ECG monitoring in textiles. Sensors and Actuators;A: Physical. 2006;**2006**:48-53. DOI: 10.1016/j.sna.2005.10.052

[5] Patel S, Park H, Bonato P, Chan L, Rodgers M. A review of wearable sensors and systems with application in rehabilitation. Journal of NeuroEngineering and Rehabilitation. 2012;**9**:21. DOI: 10.1186/1743-0003-9-21

[6] Sung M, Marci C, Pentland A. Wearable feedback systems for rehabilitation. Journal of NeuroEngineering and Rehabilitation. 2005;**2**:17. DOI: 10.1186/1743-0003-2-17

[7] Curone D, Secco EL, Tognetti A, Loriga G, Dudnik G, Risatti M, Whyte R, Bonfiglio A, Magenes G. Smart garments for emergency operators: The ProeTEX project. In: IEEE Transactions on Information Technology in Biomedicine. 2010;**14**(3):694-701

[8] Starner T.E. Powerful change. Part I. Batteries and possible alternatives for the mobile market. In: IEEE Pervasive Computing. 2003;**2**(4):86-88

[9] Corbishley P, Rodriguez-Villegas E. Breathing detection: Towards a miniaturized, wearable, battery-operated monitoring system. IEEE Transactions on Biomedical Engineering. 2008;**55**(1):196-204. DOI: 10.1109/TBME.2007.910679

[10] Schubert MB, Werner JH. Flexible solar cells for clothing. Materials Today. 2006;**9**(6): 42-50. DOI: 10.1016/S1369-7021(06)71542-5

[11] Bedeloglu A. Progress in organic photovoltaic Fibers research. In: Kosyachenko LA, editor. Solar Cells - New Aspects and Solutions. Rijeka: InTech; 2011. pp. 255-286. DOI: 10.5772/21013

[12] Xi H, Luo L, Fraisse G. Development and applications of solar-based thermoelectric technologies. Renewable and Sustainable Energy Rewievs. 2007;**11**(5):923-936. DOI: 10.1016/j.rser.2005.06.008

[13] Wischke M, Masur M, Kroener M, Woias P. Vibration harvesting in traffic tunnels to power wireless sensor nodes. Smart Materials and Structures. 2011;**20**(8):085014-085022

[14] Saha CR, O'Donnell T, Wang N, McCloskey P. Electromagnetic generator for harvesting energy from human motion. Sensors and Actuators;A: Physical. 2008;**147**(1):248-253. DOI: 10.1016/j.sna.2008.03.008

[15] Poulin G, Sarraute E, Costa F. Generation of electrical energy for portable devices. Comparative study of an electromagnetic and a piezoelectric system. Sensors and Actuators;A: Physical. 2004;**116**(3):461-471. DOI: 10.1016/j.sna.2004.05.013

[16] Shenck NS, Paradiso JA. Energy scavenging with shoe-mounted piezoelectrics. IEEE Micro. 2001;**21**(3):30-42. DOI: 10.1109/40.928763

[17] Hayakawa M. Electric wristwatch with generator. US Patent 5001 685. Mar. 1991

[18] Moss SD, McLeod JE, Powlesland IG, Galea SC. A bi-axial magnetoelectric vibration energy harvester. Sensors and Actuators;A: Physical. 2012;**175**:165-168. DOI: 10.1016/j.sna.2011.12.023

[19] Blums J, Terlecka G, Vilumsone A. The Electrodynamic human motion energy converter with planar structure. Advanced Materials Research. 2011;**222**:36-39. DOI: 10.4028/www.scientific.net/AMR.222.36

[20] Cho H-S, Yang J-H, Park S-H, Yun K-S, Kim Y-J, Lee J-H. An exploratory study on the feasibility of a foot gear type energy harvester using a textile coil inductor. Journal of Electrical Engineering & Technology. 2016;**11**(5):1210-1215. DOI: 10.5371/JEET.2016.11.5.1210

[21] Roh J-S, Chi Y-S, Lee J-H, Nam S, Kang TJ. Characterization of embroidered inductors. Smart Materials and Structures. 2010;**19**(11):115020-115032. DOI: 10.1088/0964-1726/19/11/115020

[22] Terlecka G, Viļumsone A, Blūms J. The electrodynamic human motion energy harvester in smart clothes. In: 150 Years of Research and Innovation in Textile Science: Book of Proceedings (2). 2011. pp. 866-870

[23] Kishi M, et al. Micro thermoelectric modules and their application to wristwatches as an energy source. In: Eighteenth International Conference on Thermoelectrics. IEEE; 1999. pp. 301-307. DOI: 10.1109/ICT.1999.843389

[24] Baumgartner WR. Energy system for electronic watch. U.S. Patent No. 4,320,477. 16 Mar. 1982

[25] Blums J, Terlecka G, Gornevs I, Vilumsone A. Flat inductors for human motion energy harvesting. In: Procceding SPIE. 8763, Smart Sensors, Actuators, and MEMS VI 87631L; 2013, Grenoble, France. DOI: 10.1117/12.2016995

[26] Dāboliņa I, Blūms J. Plakano induktīvo elementu dizains. RTU zinātniskie raksti. 2011, 9; Materiālzinātne (6):105-110

[27] Eglīte L, Terļecka G, Blūms J. Energy Generating Outerwear. Material Science. Textile and Clothing Technology. 2015;**10**:67-71. DOI: 10.7250/mstct.2015.010

[28] Dziurdzia P, Stepien J. Autonomous wireless link powered with harvested heat energy. IEEE International Conference on Microwaves, Communications, Antennas and Electronics Systems (COMCAS). IEEE; 2011. DOI: 10.1109/COMCAS.2011.6105890

[29] Carmo JP, Gonçalves LM, Correia JH. Thermoelectric microconverter for energy harvesting systems. IEEE Transactions on Industrial Electronics. 2010;**57**(3):861-867. DOI: 10.1109/TIE.2009.2034686

[30] Wong H-P, Zuraini D. Human body parts heat energy harvesting using thermoelectric module. In: IEEE Conference on Energy Conversion (CENCON). IEEE; 2015. p. 211-214. DOI: 10.1109/CENCON.2015.7409541

[31] Zhao X, et al. Towards low-power wearable wireless sensors for molecular biomarker and physiological signal monitoring. In: IEEE International Symposium on Circuits and Systems (ISCAS). IEEE; 2017. DOI: 10.1109/ISCAS.2017.8050558

[32] Atmel Corporation White Paper 7926A–AVR. Saether K, Fredriksen I. Introducing a new breed of microcontrollers for 8/16-bit applications. [Internet]. 2008. Available from: http://ww1.microchip.com/downloads/en/DeviceDoc/doc7926.pdf [Accessed: 15 October 2017]

[33] Rebel G, Estevez F, Gloesekoetter P, Castillo-Secilla JM. Energy harvesting on human bodies. In: A. Holzinger et al. editor, Smart Health: Open Problems and Future Challenges. LNCS 8700, pp. 125-159. 2015. Switzerland: Springer international publishing; 2015. DOI: 10.1007/978-3-319-16226-3_6

[34] Zhang Q, Wang Y, Kim ES. Electromagnetic energy harvester with flexible coils and magnetic spring for 1-10 Hz resonance. Journal of Micromechanical Systems. 2015;**24**(4): 1193-1206. DOI: 10.1109/JMEMS.2015.2393911

Chapter 4

Dielectric Elastomers for Energy Harvesting

Gordon Thomson, Daniil Yurchenko and
Dimitri V. Val

Additional information is available at the end of the chapter

http://dx.doi.org/10.5772/intechopen.74136

Abstract

Dielectric elastomers are a type of electroactive polymers that can be conveniently used as sensors, actuators or energy harvesters and the latter is the focus of this review. The relatively high number of publications devoted to dielectric elastomers in recent years is a direct reflection of their diversity, applicability as well as nontrivial electrical and mechanical properties. This chapter provides a review of fundamental mechanical and electrical properties of dielectric elastomers and up-to-date information regarding new developments of this technology and it's potential applications for energy harvesting from various vibration sources explored over the past decade.

Keywords: dielectric elastomer, circuits, materials, applications, electrostatic

1. Introduction

Energy harvesting (EH) is a process of converting energy existing in other forms or wasted energy into electrical energy that can be used as an alternative to the existing energy sources. Despite the semantics behind the words *harvesting* and *scavenging* the first one has been widely accepted in literature and will be used throughout this paper. EH from vibrations has formed its own niche because many natural phenomena as well as man-made machines and structures generate vibrations that can be converted into electrical energy. For instance, different methods can be used, various materials can be employed, and linear, parametric or nonlinear systems can be utilised for optimising the devices' efficiency. A number of transduction methods capable of such a conversion have been suggested, including piezoelectric, electromagnetic, electrostatic and triboelectric.

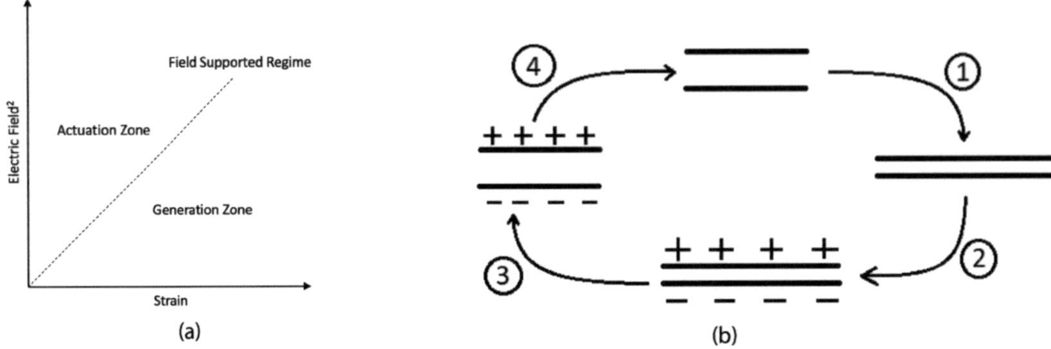

Figure 1. Basic principals of dielectric elastomer energy harvesting. (a) Actuation and generation zones. (b) The energy harvesting cycle for a dielectric elastomer.

EH using dielectric elastomeric materials is based on the principle of varying capacitance, however, instead of a traditional capacitor with two parallel plates separated by air a dielectric elastomer (DE) is used. This polymer, coated with compliant electrodes on each side, becomes a variable capacitance capacitor and can be used in two opposite modes: actuation, with a high potential in soft robotics and energy generation (**Figure 1(a)**). When the electric field is above its supported regime a DE is in the actuation zone, below—the generation one. Although there are many interesting and useful applications of DE actuators, this chapter will concentrate on the energy harvesting potential of DEs.

In the energy harvesting mode DEs can generate electricity when stretched by an externally applied force. The DE EH approach has recently seen an increasing amount of attention from a variety of science branches, including chemistry, electrical and mechanical engineering, robotics and MEMS, due to the multidisciplinary nature of the topic and a wide range of potential applications.

Pioneering work on DE EH was conducted by Pelrine et al. [1], who set the basic principles of using DEs for EH. Since then various advantages and disadvantages of this EH method have been investigated, in particular in the context of its high energy density [2]. The maximum energy density that has been achieved by a DE generator is $3.8\mu W/mm^3$ [3], whereas the electromagnetic and piezoelectric generation resulted in $2.21\,\mu W/mm^3$ [4] and $0.375\mu W/mm^3$ [5] respectively.

This chapter focuses on the DE EH applications and provides an overview of the work done up-to-date in the field, mainly after the book published in 2008 [6]. The chapter is divided into five sections based on the main components of the DE EH theory and cover electrical and material background, failure modes, initial voltage issue and applications.

2. Electrical background

The basic equation describing the behaviour of an ideal capacitor, with two parallel plates and air between them, is well-known and states that:

$$Q = CV, \quad C = \frac{\epsilon\epsilon_0 A}{z} \tag{1}$$

where Q is the charge held on the electrodes, V is the voltage over the capacitor and C is the capacitance, ϵ is the relative permittivity of the substance between the plates, $\epsilon_0 = 8.854 \times 10^{-12}\,\mathrm{Fm}^{-1}$ is the permittivity of free space, A represents the effective area of the plates and z is the distance between the plates. When the distance between the plates is changed (the plates area is assumed to be a constant) the capacitance will be changed accordingly. Thus, the variable capacitance indicates the two energy harvesting schemes: constant voltage and constant charge. First one assumes the application of an initial voltage to the plates and then the distance between the two plates is reduced, leading to an increase in the capacitance and thereby in the charge increase that can be harvested at the minimum distance between two plates. A similar approach can be employed when the constant charge scheme is used. In this case increasing the distance between the two plates reduces the capacitance and thereby increases the voltage that can be harvested. The energy gain is related to the difference in its values between the maximum and minimum capacitances. The electrical energy stored in a classical capacitor is given by the following equation:

$$U = \frac{1}{2}CV^2 = \frac{1}{2}QV \tag{2}$$

where the second equality can be obtained by using Eq. (1). However, in some applications moving physically capacitor plates is not feasible or efficient. Thus, a material mimicking the properties of a capacitor can be used instead and DEs are a perfect fit for that. Indeed, DEs are extremely poor conductors, relatively inexpensive in production and can be repeatedly deformed and stretched. To manufacture a DE-based capacitor a DE is sandwiched between two deformable electrodes serving as an insulator. Another important property of DEs from the EH point of view is their high deformability because it enables the capacitor to reach high values of the capacitance by stretching the DE and, subsequently, significantly reducing its thickness, i.e. z, as can be seen from Eq. (1) expressed in a slightly different form:

$$C = \frac{\epsilon_d\epsilon_0 A}{z} = \frac{\epsilon_d\epsilon_0 P}{z^2} \tag{3}$$

where $P = Az$ is the material volume and ε_d is the relative permittivity of the dielectric elastomer. It should be noted that whilst the area of the electrodes (plates) for a conventional electrostatic energy harvester remains constant, the area of the DE capacitor changes under deformation. Since a DE can usually be treated as an incompressible material, i.e. its volume does not change under deformation, Eq. (1) can be expressed in the form of Eq. (3). Thus, cyclic stretching can be used to increase an initial bias voltage of the capacitor to a higher voltage that can be harvested. The diagram demonstrating all stages of this repeatable EH process is presented in **Figure 1(b)**, where the numbers 1–4 in **Figure 1(b)** represent the stages of the energy harvesting cycle and may be explained as following:

1. The DE is stretched increasing the area of the electrodes and decreasing the thickness between the electrodes, leading to an increase in the capacitance.

2. A charge is placed over the DE.
3. The DE is allowed to return to its original shape that decreases the capacitance, resulting in a higher potential voltage over the elastomer.
4. The gained electrical energy is then removed.

Pelrine et al. [1] derived the following formula for estimating the energy gain per cycle, e_g:

$$e_g = \frac{P\epsilon_d\epsilon_0}{2}\left({E_c}^2 - {E_s}^2\right) \tag{4}$$

where E_c and E_s, defined as $E = V/z$, represent the electric field at the contracted and stretched states, respectively.

Having outlined the basic principle of energy harvesting using DEs, let us consider electrical methodology designed for EH. Currently, three EH schemes are being used: constant voltage, constant charge and constant field. The term 'constant' refers to the electrical state of the DE during the transition of the material from the stretched to its relaxed state. The first two schemes are the most investigated ones and have been used before in electrostatic EH; their advantages and disadvantages have been discussed in [7, 8]. The cycles' stages for constant voltage and constant charge are shown in **Figure 2(a)** and **(b)**, where λ_{min} and λ_{max} denote the minimum and maximum stretching states, respectively.

The constant voltage scheme in **Figure 2(a)** begins with stretching the DE and charging it at its maximum stretching state (line 1). Reaching the maximum voltage, controlled by the electrical breakdown (Section 5), the material is released and returns to its unstretched state along the constant voltage line 2 with the corresponding decrease in the capacitance. To maintain the constant voltage the charge needs to be decreased from Q_{high} to Q_{low}; the charge difference flows to a reservoir (e.g. battery) through the circuit controlled by diodes. The capacitor is then discharged into the reservoir so that the DE returns to its original state (line 3), and after that the next cycle can begin. The amount of harvested energy per one cycle is represented by the area circumscribed by the three lines and can be calculated as [8]:

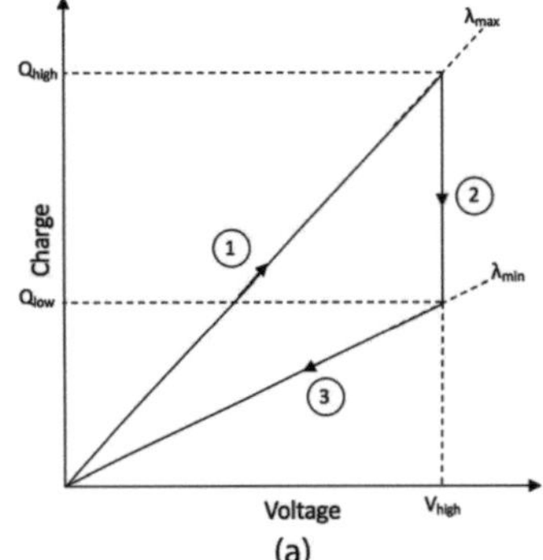

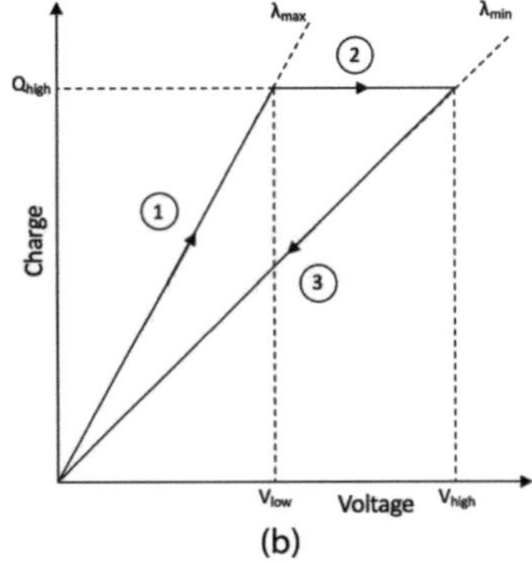

Figure 2. Common energy harvesting schemes. (a) Constant voltage scheme. (b) Constant charge scheme.

$$e_g^V = \frac{1}{2} \Delta C V_{high}^2 \tag{5}$$

where ΔC is the difference between high and low capacitance values.

A range of electrical circuits for DE generators is discussed in [7]. In particular, it is mentioned that the constant voltage scheme is the most practical one to realise. This is due to the lack of switches that can be difficult to effectively integrate into a EH device, especially considering the exact timings necessary to achieve the maximum efficiency.

The constant charge scheme is depicted in **Figure 2(b)** and uses the variable capacitor as a pump to move a charge from a low to high voltage. This is done by initially stretching the DE and placing a charge onto it while the latter is at its high stretching state (line 1). The power supply is then disconnected and the DE is released to return to its original state increasing the voltage from V_{low} to V_{high} due to the decreasing capacitance (line 2). At this point the electrical energy is harvested and the DE reverts back to its initial state (line 3). The energy harvested per one cycle is represented by the area enclosed by the three lines and equals [8]:

$$e_g^Q = \frac{1}{2} \Delta C V_{low} V_{high} \tag{6}$$

Although the constant charge scheme may be useful and effective as shown by Blokhina et al. [9], who considered a constant charge circuit for a nonlinear electrostatic vibration energy harvester, it is hampered by its need for switches which can reduce the power output and increase the complexity of the entire circuit. Thus, following this argument and discussion in [7] it is save to say that the constant voltage scheme is more preferable for implementation. However, high voltage is not suitable for most micro-power devices operating at a low voltage, in which a DC-DC voltage step down is normally needed. This extra component can reduce the overall efficiency of the DE EH so that the constant charge scheme may provide a greater efficiency for such applications.

Most recent works have been based on the electrical circuitry first implemented by Huang et al. [10], which operates according to the constant voltage scheme. In the latter work the authors devised an electrical circuit, shown in **Figure 3(a)**, which can measure the charging and harvesting currents as well as the voltage over the DE. It has also been confirmed that

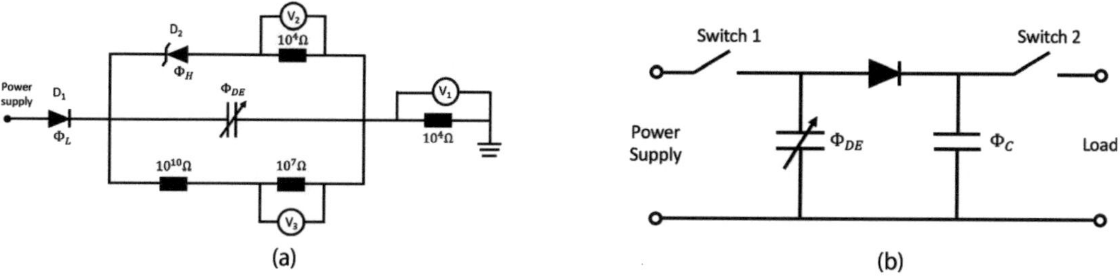

Figure 3. Dielectric elastomer energy harvesting circuits. (a) A common circuit for dielectric elastomer EH research. (b) A circuit highlighting features for the optimised scheme.

there is a proportional quadratic dependence of the dielectric capacitance on the membrane stretch. Using this result it was found that the dielectric constant of the common DE VHB 4905 (3M) was $\epsilon = 4.03 \pm 0.17$.

The circuit shown in **Figure 3(a)** works as following. The DE is charged using a power supply; the charge passes through a rectifier diode, D_1, at the maximum stretch of the DE. As the DE stretch reduces, the voltage initially increases at a constant charge, with no new charges flowing through D_1. This continues until the voltage over the DE reaches a pre-determined discharge voltage, governed by the Zener diode D_2. Once this voltage has been reached, the charge leaves the circuit through this diode. Finally, the membrane is stretched out again, lowering the voltage of the charge that remains, until the voltage reaches that of the power supply, which is connected again and the cycle is repeated. The circuit shown in **Figure 3(a)** is very useful for DE EH research since through its various components a large amount of information can be gathered about the behaviour of the DE generator. For example, using the lower of the three parallel branches, which behaves as a voltage divider, the voltage over the DE can be constantly monitored using V_3. Similarly, V_1 and V_2 can be used to measure the charging and harvesting currents, respectively, utilising the measured voltage in each voltmeter and the known resistances. Once these currents are known the input and harvesting charges can be found using $Q = \int idt$. The net energy gain can then be estimated as $\Delta E = (V_{high} - V_{low})\Delta Q$.

The above discussion is related to the ideal material behaviour with no failure modes considered and no losses taken into account. Electromechanical coupling, e.g. further stretching of a DE when a charge is applied and the corresponding change of the capacitance, has also not been mentioned. Since DEs may experience electrical and mechanical failures (Section 4) it is essential to account for them [2]. These failure modes, shown as curves on the $Q - V$ plane in **Figure 4(a)**, change the idealistic EH picture. Apparently, the maximum harvested energy corresponds to the maximum area enclosed by the lines connecting the maximum and minimum stretch states, without crossing to the failure regions. In **Figure 4(a)** such a cycle is constrained between the aforementioned upper and lower stretch states. It is also limited by two failure modes, dielectric breakdown (DB) and electromechanical instability (EMI), which are discussed in Section 4 in greater detail. An 'optimal' scheme, developed by Shian et al. [11], built upon the conventional EH schemes as shown in **Figure 4(a)**. In [11] the authors proposed

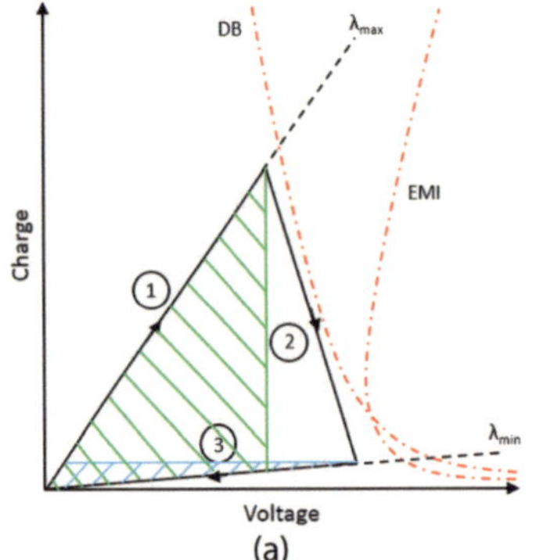

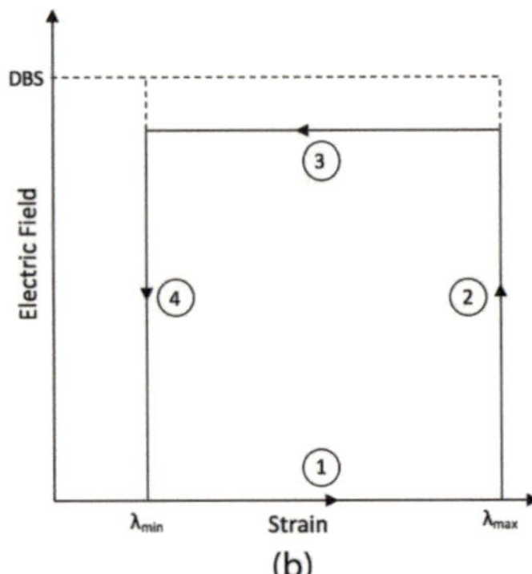

Figure 4. More recent energy harvesting schemes. (a) The optimised scheme. (b) Constant field scheme.

this new harvesting cycle, which utilises the circuit diagram shown in **Figure 3(b)**. This scheme maximises the working area within the limits imposed by the failure criteria set out in [2]. The optimal cycle is shown in **Figure 4(a)** in black, whereas the areas for the constant voltage (green) and constant charge (blue) are also superimposed into the graphs to demonstrate the difference between the schemes. One can see clearly the benefits of the optimisation by simply comparing the corresponding areas. Following one of the conventional methods (constant voltage or charge) a designer must lose large sections of the available energy in order to avoid failure. It should be noted that the values for the stretch during the cycle and the maximum charge or voltage used in the conventional schemes can be used to alter their respective areas on **Figure 4(a)**, in accordance with Eqs. (5) and (6). However, this system is hard to implement for a real DE material as its rate of the stretch decrease will not be as ideal as indicated in [12]. In that work the authors attempted to optimise the system of harvesting by showing an analytical method of determining the ideal stretch ratio at which the membrane discharging should occur. Additionally, Hoffstadt et al. [13] conducted an optimisation of the conventional EH cycles and determined the ideal charging and discharging times for maximal energy gain in the optimised scheme.

The circuit, presented in **Figure 3(b)** was proposed to work with the above EH scheme. The first phase of this cycle is charging the DE at its maximum stretch state, after which switch 1 is used to disconnect the DE from the supply. The second phase of the cycle is decreasing the DE stretching. This sees the DE discharge into a temporary storage capacitor, Φ_C, which is connected in parallel to the DE and the harvesting section of the circuit. This discharging to the capacitor will cause the charges on the actual DE to drop, however the voltage will continue to increase as shown by Eq. (3). The final stage of this cycle occurs at the low stretch state, where all electrical energy is removed from both the DE and the temporary capacitor.

A relatively new constant field EH scheme, shown in **Figure 4(b)**, proposed to improve the conventional schemes. This cycle begins with initial stretching of the DE to its maximum level (line 1). The DE is then charged up to an electric field pre-determined by the designer (line 2). After that the DE streching is reduced back to its original level at a constant electric field (line 3). It is worth to recall that the electric field is defined as $E = V/z$, therefore, at constant electric field a decrease in the DE stretching and the corresponding increase in the DE thickness lead to an increase in the voltage over the DE. This is an advantage over the constant charge scheme, since at the latter the electric field increases during the reduction of DE stretching that may cause failure due to dielectric breakdown (Section 4). The final stage of the circuit is to remove the electrical energy from the DE, returning the latter to its original state (line 4). Since failure by dielectric breakdown is of no concern for this scheme this allows the DE to be operated closer to its dielectric breakdown strength (Section 4). Czech et al. [14] analytically compared the ideal constant voltage, constant charge and constant electric field schemes and found that the constant electric field was the most effective. However, it should be noted that maximising the energy output by increasing the electric field may have an adverse effect on the lifespan of the generator [15].

A few other works have been recently done on improving electrical aspects of DE EH, e.g. Kaltseis et al. [16] developed an experiment based on operating a DE generator between two charged reservoirs of different voltages that allowed them to monitor separately electrical and mechanical energies and determine the net energy gain.

3. Material background

DEs belong to a group of electroactive polymers (EAP) that received this name because of their ability of changing their shape under applied electrical field. Essential properties of DEs for EH applications are low mechanical stiffness, high dielectric constant, high electrical breakdown strength, high elastic energy density and high deformability. There are a number of DEs possessing these properties including acrylates (e.g. VHB 4905 and 4910), silicones, polyurethanes, rubbers, etc. It is also worth to note that DEs have a number of properties such as nonlinear stress-strain relationship, viscoelasticity (i.e. time-dependent behaviour), electromechanical coupling, temperature-dependent performance, etc., which significantly complicate the modelling and design of DE generators. Quite extensive overview of the main properties of these materials in the context of EH can be found in [17–21]. The relevant material properties of DEs and their modelling are also briefly considered further in this section.

3.1. Material properties

A typical DE-based system (i.e. capacitor) comprises a DE membrane sandwiched between two electrodes, as depicted in **Figure 5**. Depending on the purpose of the system in-plane stretching of the DE can be done either by applying an electrical field (actuator) or mechanical deformation (harvester). The applied electrical field causes the so-called Maxwell stress in the material due to the interaction of opposite charges on both electrodes, resulting in the DE stretching. This concept has been well demonstrated, e.g. by Mockensturm et al. [22] who manipulated a DE membrane by altering the applied electric field to move the membrane between its various steady state equilibria. Tensile stresses in the plane of the DE membrane, caused by in-plane or out-of-plane loading, lead to the material deformation reducing the membrane thickness and increasing its area, thereby changing the capacitance of the DE-based capacitor.

A well-known material property of DEs is their high deformability, i.e. the ability to undergo large deformations without rupture. The stretch ratio (or simply stretch), defined as a ratio of the stretched length to the initial one $\lambda = L_s/L$, may reach for some materials up to 900% [23, 24]. This property is very important since it enables to impose relatively high levels of

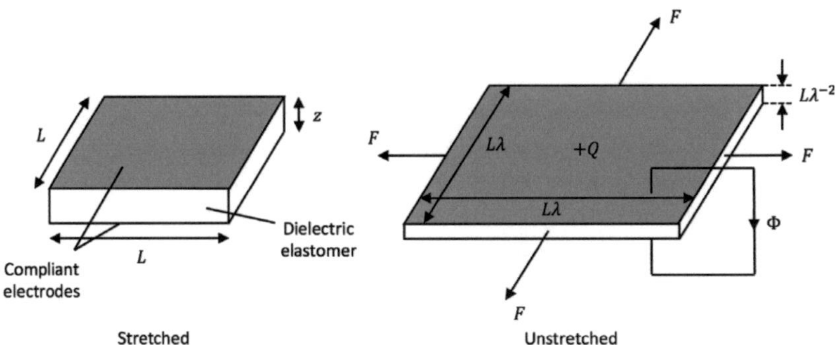

Figure 5. Loading of a dielectric elastomer.

pre-stretch on DE membranes that improves their material performance, in particular increases the dielectric (i.e. electrical breakdown) strength and reduces the effective compressive modulus [25]. At low deformations the material demonstrates strain softening that changes to stiffening at higher deformations. This type of mechanical behaviour is usually described by a hyperelastic constitutive model under the assumption that the DE is an ideally elastic material. Different hyperelastic models used for DEs will be briefly described in the next sub-section. To determine parameters of such models data from uniaxial tensile tests, and for more accurate determination also from biaxial tests, are needed. Results of uniaxial and biaxial tests of VHB 4910 were reported in [26, 23] and for silicone-based elastomers in [27]. Results of uniaxial tests of natural rubbers can be found in [28].

Another important property that significantly affects the mechanical behaviour of DEs is viscoelasticity, which occurs due to the realignment of polymer chains of the materials under loading and leads to time-dependence of deformations and stresses. For example, stress-stretch (or strain) curves of DEs depend on the strain (i.e. loading) rate and are usually show stiffening with increase in the rate, e.g. [24]. In particular, viscoelasticity causes dissipation, i.e. losses, of the mechanical energy that, in its turn, reduces the efficiency of DE generators [10, 29–31]. The losses are associated with a hysteretic behaviour of a DE, i.e. the stress-stretch curves of the material follow different paths during loading-unloading forming a hysteresis loop whose area represents the energy loss, e.g. [32]. Creep and stress relaxation are other phenomena associated with viscoelasticity. Creep is usually defined as a time-dependent increase in deformation under constant load, while stress relaxation is a decrease in stress under constant deformation. Creep is more important for DE actuators, while stress relaxation may have a significance influence on the performance of both DE generators and actuators. Typical tests to determine viscoelastic characteristics of a DE as well as parameters of a model describing the material's viscoelastic behaviour include loading-unloading tests at a range of strain rates, single-step relaxation tests at various stretch levels and multi-step relaxation tests, in which at each step loading follows by a sufficiently long holding period to allow for stress relaxation [32]. The tests are usually uniaxial, however, equi-biaxial tests may also be useful. Most of the experimental studies on the viscoelastic behaviour of DEs have been carried out for VHB 4905 and VHB 4910 [24, 26, 32–36]. In particular, it has been shown that for this material very rapid stress relaxation occurs within the first few seconds of holding, while after 30 min the remaining stress almost stabilises [32]. Past experiments also showed that for VHB polymers the viscoelastic relaxation time is of the order of 10^2 seconds at room temperature [37]. Pharr et al. [24] demonstrated that the strain rate and the specimen size had a noticeable influence on rupture of VHB 4905. Hossain et al. [33] reported results of relaxation tests under both mechanical and electric loading and showed that the latter had a major effect on the time-dependent behaviour of VHB 4910. A few results of similar experiments for other DEs have been published. In particular, Schmidt et al. [23] presented results of uniaxial and equi-biaxial relaxation tests for Interpenetrating Polymer Network Reinforced Acrylic Elastomers (IPN), which were produced on the basis of VHB 4910. Bernardi et al. [27] published similar results for silicone-based DEs.

A hysteretic behaviour of DEs, especially of rubbers and rubber-like materials, and associated with it loss of mechanical energy can also be due to the so-called Mullins effect, also known as stress softening [38]. This effect causes a decay of the elastic stiffness of a material under repeated loading-unloading, which depends on the maximum stretch previously experienced by the material. Recommendations regarding experimental characterisation of the effect can be found in [39].

Effects of the operating temperature on the material properties of VHB 4910 and the performance of a generator made of this material were investigated by Chen et al. [40]. In particular, it was found that the generator operated more efficiently under lower temperatures due to smaller viscoelastic losses.

As can be seen from above, so far VHB 4905/10 (3M^{TM}) have received most attention from researchers. However, other DEs, in particular natural rubbers and polyurethanes (PUR), have been also investigated and demonstrated good potential for EH, in some cases better than VHB, e.g. [28, 41, 42]. One of the researches, [41], examined the influence of stretching on the electrical properties, such as the electrical (or dielectric) breakdown strength (DBS) and dielectric constant, of VHB 4910 and natural rubber and suggested that the latter had advantages over VHB in the context of EH. It also necessary to note that in our overview of the DE material properties we mainly concentrated on their mechanical properties, while the electrical properties were only mentioned in relation to the last referred publication. Of course, the electrical properties of DEs are also very important for EH and were considered in a number of the references provided above. Detailed recommendations on experiments that can be used to determine both mechanical and electrical properties of DEs are given in [39].

Finally, based on data presented in [43] we include **Table 1**, which provides the important mechanical and electrical properties of some DEs that can be used for EH. More information about specific DEs appearing in the table, in particular polyurethanes (i.e. PUR 1, PUR 2, etc.), can be found in the source [43]; acrylate means VHB 4905.

3.2. Modelling

Modelling is essential for the development and design of DE-based devices for EH. The idea is to simulate the performance of such devices using analytical and/or numerical models and reduce the need for time-consuming and relatively expensive experiments. Basic theory of DEs

Name	Mechanical properties				Electrical properties	
	DE thickness (μm)	Break Strain (%)	E at 50% (MPa)	Creep (30 min) (10% Def.)	ε_d (1/8 Hz)	DBS (MV/m)
PUR 1	50	421	3.36	24	6	218
PUR 2	50	252	1.47	2.9	7.1	108
PUR 3-1	49	251	1.96	1.9	7.1	132
PUR 3-2	97	319	1.3	6.5	8.0	102
Silicone	45	422	0.25	16	2.5	80
Acrylate	498	879	0.04	70	4.5	31

Table 1. Comparison of various elastomer properties.

based on thermodynamics and continuum mechanics was described in [44]. In this section we overview some specifics of its practical implementation.

As has been noted previously, the elastic behaviour of DEs under large finite deformations is usually modelled within the hyperelasticity framework. This means that the material is assumed to be ideally elastic and that stress-strain (or stretch) relationships are obtained based on a strain energy density function, which is expressed as a function of the invariants of the Cauchy-Green deformation tensor, e.g. [45]. Thus, a particular hyperelastic constitutive model fully depends on the corresponding formulation of the strain energy density function. So far various formations of the strain energy density function and, subsequently, hyperelastic models have been used to describe the elastic behaviour of DEs under large deformations, e.g. Neo-Hookean [46–48], Mooney-Rivlin [49], Ogden [27, 49, 50], Yeoh [51], Arruda-Boyce [26, 52] and Gent [48, 53, 54]. Of course, this list is far from being exhaustive. There are other hyperelastic models that could also be used for modelling the constitutive behaviour of DEs. Many researchers have compared the predictions provided by different hyperelastic models with the experimentally observed behaviour of DE membranes, e.g. Yeoh, Ogden and Arruda-Boyce [26], Mooney-Rivlin and Ogden [55], Neo-Hookean and Gent [48]. It is important to note there is no universal hyperelastic model that would provide accurate and reliable predictions for all DEs. A hyperelastic model that yields excellent results for a particular DE under specific conditions may be inapplicable for another DE. An extensive review of the hyperelastic models can be found in [56, 57].

The hyperelastic models describe the time-invariant behaviour of DEs. However, this is not the case because of viscoelasticity. Thus, in order to achieve more realistic predictions of the DE performance under mechanical and/or electrical loading, especially cyclic or dynamic loading, the time-dependency due to viscoelasticity needs to be taken into account. Wissler and Mazza [51] suggested to use the quasi-linear viscoelastic (QLV) model. This model is based on the assumption that the relaxation function is independent of the stretch ratio (or strain), i.e. depends only on time. The relaxation function was determined based on experimental results using the so-called Prony series. To combine the hyperelastic and QLV models, the coefficients of the energy density function are multiplied by the relaxation function. The same approach was also used by Wang et al. [58]. The model provides reasonable for quasi-static creep and stress relaxation problems; however, its predictions noticeably deviated from experimental results in the case of dynamic loading as the number of cycles increased [26].

Many researchers use models based on classical rheological models such as Maxwell and Kelvin-Voight with added elements. One of the most popular models is the so-called standard linear solid model presents a Maxwell element in parallel with an elastic spring, e.g. [37, 59, 60]. Others used an extended Kelvin-Voight model, i.e. a Kelvin-Voight element in series with an elastic spring [58, 61]. A model combining in parallel Maxwell and Kelvin-Voight elements has also been suggested [62]. Models based on the generalised Maxwell model, also known as Maxwell-Wiechert model, have also been proposed and implemented [56, 63, 64]. An extensive review of viscoelastic models can be found in [65].

It is worth to note that the viscoelastic models described above were usually implemented for uniaxial or equi-biaxial problems that led to a set of ordinary differential equations, which were solved numerically. However, to solve more general problems or obtain more accurate results (i.e. avoid simplifying assumptions) the finite element (FE) modelling is needed. One of the main challenges here is to develop and implement finite elements that take into account electromechanical coupling since commercial FE software, e.g. ABAQUS, does not currently provide such an option. FE models for simulating the DE behaviour under both mechanical and electrical loads have been proposed in [66–68]. In particular, Henann et al. [67] employed the developed FE model to carry out an analysis of both DE actuator and generator. Jean-Mistral et al. [69] also used FE modelling to investigate a new technique for DE EH.

4. Failure modes

In order to fully asses the DE generators potential the failure mechanisms have to be considered. In general there are two types of failures a DE device may face: material and manufacturing related. Whereas the former are discussed in this section the latter are not because they are mostly related to imperfections due to the manufacturing process. This group includes but not limited to failures due to non-uniform stretching of the material over the working area, poor or uneven distribution of charges (electrodes) over the working surface, loss of contact between electrodes and working surfaces, others malfunctioning of electrodes during normal operation. The material related failures may be combined into four groups: rupture by stretch, loss of tension, electrical breakdown and electromechanical instability.

Rupture by stretching occurs when the membrane is stretched too much and the internal strain becomes too high leading to the material ripping, as explained by the classical stress-strain curve for a material. Loss of tension occurs when the initial voltage placed over the membrane is enough to remove the pre-strain put onto the generator in its construction.

Dielectric breakdown is a complex topic, however it is enough to say that there is a specific electric field at which the membrane will essentially become a conductor and the generator will fail. As the electric field is given by $E = V/z$, the dielectric breakdown strength (DBS) is inversely proportional to the thickness, thus the thinner the material the larger its DBS [70]. This was investigated further by Tröls et al. [41] who reported how the dielectric breakdown strength and dielectric constant is influenced by the stretch of the material. Their main findings regarding E_{EB} and ϵ for 3M^{TM} VHB 4910 are shown in **Table 2**. Pre-straining, typically performed during the manufacturing stage, can improve the DBS. The pre-straining is done by taking the DE to a minimum level of stretch λ_{min} in a device. Besides improving the DBS of the material, pre-straining negates the risk of failure by loss of tension. The increase in DBS of VHB 4910 (3M^{TM}) by 1100% under an equal biaxial pre-straining of 500% has also been reported in [71].

Value	$\lambda = 1$	$\lambda = 2$	$\lambda = 3$	$\lambda = 4$	$\lambda = 5$
E_{EB} Vμm^{-1}	–	100	124	145	163
ϵ (1 Hz)	4.24	–	3.83	–	3.44

Table 2. Dependence of the electrical properties of VHB 4910 on the stretch ratio.

Electromechanical instability occurs due to nonlinearities in the materials behaviour. As the membrane stretches due to increasing electrostatic forces, nonlinearities occur in the stretching and some areas will become wrinkled producing thinner and thicker sections.

The failure modes associated with dielectric elastomers were investigated by Koh et al. [2] who derived the equations of failure for the two work conjugate planes of force-displacement and voltage-charge. These equations can then be plotted as lines on the respective graphs with the enclosed area representing the maximum working area, which can be used to find the maximum energy per cycle for a specific membrane. This work showed that for a membrane with modulus, $\mu = 10^6 \mathrm{N/m^2}$, permittivity of $\epsilon = 3.54 \times 10^{-11} \mathrm{F/m}$ and density of $\rho = 1000\ \mathrm{kg/m^3}$, an energy gain of 6.3 J/g per cycle was achieved. However, this would be difficult to implement in experimental conditions. Pelrine and Kornbluh [72] then updated this describing the fundamental science and derivations behind dielectric elastomers as well as covering the failure methods and a variety of dielectric elastomer configurations. In addition to the commonly cited failure limits set out by Koh et al. [2], Zhou et al. [73] conducted an investigation into the performance and lifetime of dielectric elastomer generators subjected to a cyclic deformation. Their work hypothesised that a decrease in the generator output is due to fatigue cracks forming on the membrane. In order to minimise this effect a smaller strain and deformation rate is proposed to increase the lifetime of the device. With an increase in applied voltage keeping the output energy higher, in accordance with Eq. (4).

Other failures that have not been mentioned above may be associated with the loss of electrical contacts, different stretching capabilities of the DEs and electrodes, fatigue and durability, as well as multiscale instabilities [74, 75].

5. Initial voltage issue

One of main concerns with dielectric elastomer generators (DEGs) is the need for an initial voltage. Given that the larger the supply voltage, the larger the energy gain, ideally the supply would be in the kilovolt range, which causes an issue for small-scale or portable generators. One solution to this problem is self-priming circuits proposed by McKay et al. [76]. This circuit allows any energy not used in powering the load to go back into the energy supply, so that over time this increases the power input to the system. In their experiment the input voltage was taken from an initial 10 V to 3250 V after only 236 cycles. This work is the basis for many future real world applications, since a high voltage DC power supply or transformer is not commonly available at small-scale portable sizes. This work would allow generators to be started with low voltage batteries and power up to a working voltage. Illenberger et al. [77] further advanced the theoretical aspect of self-priming circuits by providing a mathematical analysis backed up by experimental validation with an accuracy of 0.1% after five cycles. Panigrahi and Mishra [78] built and simulated an electrical model of a dielectric elastomer generator using the common electrical simulation software, Pspice.

The use of self-priming circuits would allow other vibration energy harvesting techniques to be used to initialise a dielectric elastomer generator, which has been explored in some works including [69]. Their work used electret technology to generate a bias voltage of 100 V which acted as the input to the DEG. In a later publication [79] this idea was studied by a finite element modelling. In general there is a great potential in combining some other types of energy generation with the self-priming circuit to raise the bias voltage to a level required.

6. Applications

A clear metric of progress being made in any energy generation field is evidenced in the new applications and performances of new devices. For that reason this section will highlight some of the significant breakthroughs made in recent years.

6.1. Indoor tests

Huang et al. [80] developed a DEG that was stretched radially, in-plane, in a laboratory setting. Their device had an energy density of 550 J/kg and an efficiency of 22.1%. McKay et al. [81] developed and tested a generator in laboratory conditions that consisted of 42 membranes placed on top of each other. The total volume of the device was 0.86 cm^3 and it could produce an output power of $300\mu W$ at an excitation frequency of 0.5 Hz. This design was further developed [3] using 48 membranes within the same volume and a larger applied voltage saw the output power increase to 1.8 mW at 1.6 Hz. Lai et al. [82] conducted a laboratory test on dielectric elastomers in the application of human walking. Their device was simple in nature and generated a relatively low energy gain, $10 - 50\mu J$, however it did experimentally prove the validity of energy harvesting from human walking. Kaltseis et al. [16] conducted a laboratory based experiment on a dielectric elastomer generator in diaphragm mode and showed a specific energy of 102 mJ/g and an average power of 17 mW/g, however it only achieved a 7.5% efficiency.

6.2. Wave energy converters

There are several types of wave energy converters that utilises DEs. Moretti et al. [83] proposed a generator called poly-surge that has to be sealed to the seabed. Their numerical study based on this design provided an output of 1.56 MW; however, this was based on $15 m^3$ of dielectric elastomer and a model, which unrealistically assumed no losses due to viscoelasticity. Later Moretti et al. [84] developed a generator, similar in principle to an oscillating water column (OWC) generator, but using DEs instead of an air turbine. Their generator had an output of 0.3 J, or 173 J/kg, which is equivalent to 0.15 W. Vertechy et al. [85] have also investigated an OWC generator and experimentally tested it in a tank at a scale of 1:50. Their scaled model produced a maximum power of 76.8 mW. However, if the model was built at its full scale size, the authors estimated that it would be capable of producing 68 kW of power at the optimal wave conditions of 2 m height and a period of 11.7 s. Lv et al. [48] developed a DE-based wave energy converter for capturing heaving motion of waves. Their work showed that as the input voltage increased the experimental and theoretical power output both increased as expected, but the efficiency dropped. This was mainly due to the current leakage at high voltage. The approximate energy output of the system was 0.25 mJ per cycle. Binh et al. [86] designed a floating wave energy converter that uses a control system, which aided in maximising the generated energy. The proposed control system allowed to increase the device efficiency of up to 25% in numerical simulations.

6.3. Vibro-impact harvester

Recently a novel DE-based device for EH from a vibrational motion has been proposed in [87, 88]. The main idea centres around a hollow cylinder, excited by an external force, and a free moving internal ball (friction was neglected), motion of which is constrained at both the ends of the cylinder by DE membranes. The ball motion is excited through its impacts against the DE membranes, which causes the membranes deformation (increase in the capacitance) leading to energy harvesting. The cylinder can be placed under an angle β with respect to the horizon. For an inclination angle $\beta = 0$ the motion of the cylinder should be large enough to engage the ball motion, however for any angle $\beta > 0$ the ball will be at rest on the lower membrane and its separation from the membrane will be engaged under a set of certain conditions. This design was called vibro-impacting EH and the dynamics of the generator has been investigated for various initial conditions, angles of inclination and restitution coefficients under deterministic and random loading. Numerically simulations have shown the rich nonlinear dynamics of the harvesters and indicated a mechanism for the device optimisation.

7. Conclusions

This chapter covers the fundamental aspects of Dielectric Elastomers and their potential application for Energy Harvesting. The chapter intended to overview the new developments made since the publication of [6]. There has been a significant progress made in understanding the material and electrical properties of DEs, optimisation of the energy harvesting process and modelling DEs. Some new ideas of ceasing the need for a substantial initial voltage using a self-priming circuit and ferroelectrics/electrets have been explored. New concepts of DE-based devices have been designed and tested. Nevertheless, it would be wrong to say that the behaviour of DEs and their capabilities have been entirely studied, thus more investigations are required.

Author details

Gordon Thomson[1]*, Daniil Yurchenko[1] and Dimitri V. Val[2]

*Address all correspondence to: grt2@hw.ac.uk

1 IMPEE, Heriot-Watt University, UK

2 EGIS, Heriot-Watt University, UK

References

[1] Pelrine R, Kornbluh R, Eckerle J, Jeuck P, Oh S, Pei Q, Stanford S. Dielectric elastomers: Generator mode fundamentals and applications. Proceedings of SPIE. 2001;**4329**:148-156

[2] Koh SJA, Zhao X, Suo Z. Maximal energy that can be converted by a dielectric elastomer generator. Applied Physics Letters. 2009;**94**(26)

[3] Mckay TG, Rosset S, Anderson IA, Shea H. Dielectric elastomer generators that stack up. Smart Materials and Structures. 2014;**24**(1)

[4] El-Hami M, Glynne-Jones P, White NM, Hill M, Beeby S, James E, Brown AD, Ross JN. Design and fabrication of a new vibration-based electromechanical power generator. Sensors and Actuators A: Physical. 2001;**92**(1):335-342

[5] Roundy S, Wright PK, Rabaey J. A study of low level vibrations as a power source for wireless sensor nodes. Computer Communications. 2003;**26**(11):1131-1144

[6] R Pelrine, H Prahlad. Generator Mode: Devices and Applications. Oxford: Elsevier. 2008

[7] Kornbluh RD, Pelrine R, Prahlad H, Wong-Foy A, McCoy B, Kim S, Eckerle J, Low T. From Boots to Buoys: Promises and Challenges of Dielectric Elastomer Energy Harvesting. Boston, MA: Springer US; 2012. pp. 67-93

[8] Miranda JOM. Electrostatic vibration-to-electric energy conversion [PhD thesis]. Massachusetts Institute of Technology; 2004

[9] Blokhina E, Galayko D, Basset P, Feely O. Steady-state oscillations in resonant electrostatic vibration energy harvesters. IEEE Transactions on Circuits and Systems I: Regular Papers. 2013;**60**(4):875-884

[10] Huang J, Shian S, Suo Z, Clarke DR. Maximizing the energy density of dielectric elastomer generators using equi-biaxial loading. Advanced Functional Materials. 2013;**23**(40): 5056-5061

[11] Shian S, Huang J, Zhu S, Clarke DR. Optimizing the electrical energy conversion cycle of dielectric elastomer generators. Advanced Materials. 2014;**26**(38):6617-6621

[12] Zhou J, Jiang L, Khayat RE. Analysis on the energy harvesting cycle of dielectric elastomer generators for performance improvement. EPL (Europhysics Letters). 2016;**115**(2)

[13] Hoffstadt T, Graf C, Maas J. Optimization of the energy harvesting control for dielectric elastomer generators. Smart Materials and Structures. 2013;**22**(9)

[14] Czech B, van Kessel R, Bauer P, Ferreira JA, Wattez A. Energy harvesting using dielectric elastomers. In: Power Electronics and Motion Control Conference (EPE/PEMC), 2010 14th International; IEEE; 2010. pp. S4-18

[15] Kornbluh R, Wong-Foy A, Ron Pelrine, Prahlad H, McCoy B. Long-lifetime all-polymer artificial muscle transducers. MRS Online Proceedings Library Archive. 2010 Jan; 1271

[16] Kaltseis R, Keplinger C, Baumgartner R, Kaltenbrunner M, Li T, Mächler P, Schwödiauer R, Suo Z, Bauer S. Method for measuring energy generation and efficiency of dielectric elastomer generators. Applied Physics Letters. 2011;**99**(16)

[17] Jean-Mistral C, Basrour S, Chaillout JJ. Comparison of electroactive polymers for energy scavenging applications. Smart Materials and Structures. 2010;**19**(8)

[18] Kornbluh RD, Pelrine R, Prahlad H, Wong-Foy A, McCoy B, Kim S, Eckerle J, Low T. Dielectric elastomers: Stretching the capabilities of energy harvesting. MRS Bulletin. 2012; **37**(3):246-253

[19] Biggs J, Danielmeier K, Hitzbleck J, Krause J, Kridl T, Nowak S, Orselli E, Quan X, Schapeler D, Sutherland W, et al. Electroactive polymers: Developments of and perspectives for dielectric elastomers. Angewandte Chemie International Edition. 2013;**52**(36): 9409-9421

[20] Romasanta LJ, Lopez-Manchado MA, Verdejo R. Increasing the performance of dielectric elastomer actuators: A review from the materials perspective. Progress in Polymer Science. 2015;**51**:188-211

[21] Madsen FB, Daugaard AE, Hvilsted S, Skov AL. The current state of silicone-based dielectric elastomer transducers. Macromolecular Rapid Communications. 2016;**37**(5):378-413

[22] Mockensturm EM, Goulbourne N. Dynamic response of dielectric elastomers. International Journal of Non-Linear Mechanics. 2006;**41**(3):388-395

[23] Schmidt A, Rothemund P, Mazza E. Multiaxial deformation and failure of acrylic elastomer membranes. Sensors and Actuators A: Physical. 2012;**174**:133-138

[24] Pharr M, Sun JY, Suo Z. Rupture of a highly stretchable acrylic dielectric elastomer. Journal of Applied Physics. 2012;**111**(10)

[25] Jiang L, Betts A, Kennedy D, Jerrams S. Investigation into the electromechanical properties of dielectric elastomers subjected to pre-stressing. Materials Science and Engineering: C. 2015;**49**:754-760

[26] Wissler M, Mazza E. Mechanical behavior of an acrylic elastomer used in dielectric elastomer actuators. Sensors and Actuators A: Physical. 2007;**134**(2):494-504

[27] Bernardi L, Hopf R, Ferrari A, Ehret AE, Mazza E. On the large strain deformation behavior of silicone-based elastomers for biomedical applications. Polymer Testing. 2017;**58**(Supplement C):189-198

[28] Kaltseis R, Keplinger C, Koh SJA, Baumgartner R, Goh YF, Ng WH, Kogler A, Tröls A, Foo CC, Suo Z, et al. Natural rubber for sustainable high-power electrical energy generation. RSC Advances. 2014;**4**(53):27905-27913

[29] Zhang J, Wang Y, Chen H, Li B. Energy harvesting performance of viscoelastic polyacrylic dielectric elastomers. International Journal of Smart and Nano Materials. 2015;**6**(3): 162-170

[30] Zhou J, Jiang L, Khayat RE. Methods to improve harvested energy and conversion efficiency of viscoelastic dielectric elastomer generators. Journal of Applied Physics. 2017;**121**(18)

[31] Wang H, Wang C, Yuan T. On the energy conversion and efficiency of a dielectric electroactive polymer generator. Applied Physics Letters. 2012;**101**(3)

[32] Hossain M, Vu DK, Steinmann P. Experimental study and numerical modelling of vhb 4910 polymer. Computational Materials Science. 2012;**59**:65-74

[33] Hossain M, Vu DK, Steinmann P. A comprehensive characterization of the electromechanically coupled properties of vhb 4910 polymer. Archive of Applied Mechanics. 2015;**85**(4):523-537

[34] Sahu R, Patra K, Szpunar J. Experimental study and numerical modelling of creep and stress relaxation of dielectric elastomers. Strain. 2015;**51**(1):43-54

[35] Helal A, Doumit M, Shaheen R. Biaxial experimental and analytical characterization of a dielectric elastomer. Applied Physics A. 2018;**124**(1):2

[36] Zhang R, Iravani P, Keogh P. Closed loop control of force operation in a novel self-sensing dielectric elastomer actuator. Sensors and Actuators A: Physical. 2017;**264**(Supplement C):123-132

[37] Foo CC, Cai S, Jin Adrian Koh S, Bauer S, Suo Z. Model of dissipative dielectric elastomers. Journal of Applied Physics. 2012;**111**(3)

[38] Diani J, Fayolle B, Gilormini P. A review on the Mullins effect. European Polymer Journal. 2009;**45**(3):601-612

[39] Carpi F, Anderson I, Bauer S, Frediani G, Gallone G, Gei M, Graaf C, Jean-Mistral C, Kaal W, Kofod G, et al. Standards for dielectric elastomer transducers. Smart Materials and Structures. 2015;**24**(10)

[40] Chen SE, Deng L, He ZC, Li E, Li GY. Temperature effect on the performance of a dissipative dielectric elastomer generator with failure modes. Smart Materials and Structures. 2016;**25**(5)

[41] Tröls A, Kogler A, Baumgartner R, Kaltseis R, Keplinger C, Schwödiauer R, Graz I, Bauer S. Stretch dependence of the electrical breakdown strength and dielectric constant of dielectric elastomers. Smart Materials and Structures. 2013;**22**(10)

[42] Yin G, Yang Y, Song F, Renard C, Dang ZM, Shi CY, Wang D. Dielectric elastomer generator with improved energy density and conversion efficiency based on polyurethane composites. ACS Applied Materials & Interfaces. 2017;**9**(6):5237-5243

[43] Graf C, Hitzbleck J, Feller T, Clauberg K, Wagner J, Krause J, Maas J. Dielectric elastomer–based energy harvesting: Material, generator design, and optimization. Journal of Intelligent Material Systems and Structures. 2014;**25**(8):951-966

[44] Suo Z. Theory of dielectric elastomers. Acta Mechanica Solida Sinica. 2010;**23**(6):549-578

[45] Carpi F, Gei M. Predictive stress–stretch models of elastomers up to the characteristic flex. Smart Materials and Structures. 2013;**22**(10)

[46] Graf C, Aust M, Maas J, Schapeler D. Simulation model for electro active polymer generators. In: Solid Dielectrics (ICSD) 2010 10th IEEE International Conference on 2010 Jul 4. IEEE; pp. 1-4

[47] Shiju E, Ge C, Cao J, Liu A, Jin L, Jiang X. Research on power generation of dielectric elastomer based on MATLAB. In: Mechatronics and Automation (ICMA), 2015 IEEE International Conference on 2015 Aug 2. IEEE; pp. 549-554

[48] Lv X, Liu L, Liu Y, Leng J. Dielectric elastomer energy harvesting: Maximal converted energy, viscoelastic dissipation and a wave power generator. Smart Materials and Structures. 2015;**24**(11)

[49] Wissler M, Mazza E. Modeling of a pre-strained circular actuator made of dielectric elastomers. Sensors and Actuators A: Physical. 2005;**120**(1):184-192

[50] Qu S, Suo Z. A finite element method for dielectric elastomer transducers. Acta Mechanica Solida Sinica. 2012;**25**(5):459-466

[51] Wissler M, Mazza E. Modeling and simulation of dielectric elastomer actuators. Smart Materials and Structures. 2005;**14**(6)

[52] Park HS, Suo Z, Zhou J, Klein PA. A dynamic finite element method for inhomogeneous deformation and electromechanical instability of dielectric elastomer transducers. International Journal of Solids and Structures. 2012;**49**(15):2187-2194

[53] Li T, Qu S, Yang W. Energy harvesting of dielectric elastomer generators concerning inhomogeneous fields and viscoelastic deformation. Journal of Applied Physics. 2012; **112**(3)

[54] Liu J, Foo CC, Zhang ZQ. A 3d multi-field element for simulating the electromechanical coupling behavior of dielectric elastomers. Acta Mechanica Solida Sinica. 2017;**30**(4):374-389

[55] Fox JW, Goulbourne NC. On the dynamic electromechanical loading of dielectric elastomer membranes. Journal of the Mechanics and Physics of Solids. 2008;**56**(8):2669-2686

[56] Steinmann P, Hossain M, Possart G. Hyperelastic models for rubber-like materials: Consistent tangent operators and suitability for Treloar's data. Archive of Applied Mechanics. 2012;**82**(9):1183-1217

[57] Hossain M, Steinmann P. More hyperelastic models for rubber-like materials: Consistent tangent operators and comparative study. Journal of the Mechanical Behavior of Materials. 2013;**22**(1–2):27-50

[58] Wang Y, Chen H, Wang Y, Li D. A general visco-hyperelastic model for dielectric elastomers and its efficient simulation based on complex frequency representation. International Journal of Applied Mechanics. 2015;**7**(01)

[59] Zhang J, Chen H. Electromechanical performance of a viscoelastic dielectric elastomer balloon. International Journal of Smart and Nano Materials. 2014;**5**(2):76-85

[60] Kollosche M, Kofod G, Suo Z, Zhu J. Temporal evolution and instability in a viscoelastic dielectric elastomer. Journal of the Mechanics and Physics of Solids. 2015;**76**:47-64

[61] Chang M, Wang Z, Tong L, Liang W. Effect of geometric size on mechanical properties of dielectric elastomers based on an improved visco-hyperelastic film model. Smart Materials and Structures. 2017;**26**(3)

[62] Zhang J, Ru J, Chen H, Li D, Lu J. Viscoelastic creep and relaxation of dielectric elastomers characterized by a Kelvin-Voigt-Maxwell model. Applied Physics Letters. 2017;**110**(4)

[63] Hong W. Modeling viscoelastic dielectrics. Journal of the Mechanics and Physics of Solids. 2011;**59**(3):637-650

[64] Bortot E, Denzer R, Menzel A, Gei M. Analysis of viscoelastic soft dielectric elastomer generators operating in an electrical circuit. International Journal of Solids and Structures. 2016;**78**:205-215

[65] Wineman A. Nonlinear viscoelastic solids—A review. Mathematics and Mechanics of Solids. 2009;**14**(3):300-366

[66] O'Brien B, McKay TG, Calius E, Xie S, Anderson I. Finite element modelling of dielectric elastomer minimum energy structures. Applied Physics A: Materials Science & Processing. 2009;**94**(3):507-514

[67] Henann DL, Shawn AC, Bertoldi K. Modeling of dielectric elastomers: Design of actuators and energy harvesting devices. Journal of the Mechanics and Physics of Solids. 2013; **61**(10):2047-2066

[68] Cohen N, Menzel A, et al. Towards a physics-based multiscale modelling of the electromechanical coupling in electro-active polymers. In: Proceedings of the Royal Society of London. 2016 Feb 1;**472**(2186):20150462

[69] Jean-Mistral C, Vu Cong T, Sylvestre A. Advances for dielectric elastomer generators: Replacement of high voltage supply by electret. Applied Physics Letters. 2012;**101**(16)

[70] O'Dwyer JJ. The Theory of Electrical Conduction and Breakdown in Solid Dielectrics. Clarendon Press; 1973

[71] Kofod G, Sommer-Larsen P, Kornbluh R, Pelrine R. Actuation response of polyacrylate dielectric elastomers. Journal of Intelligent Material Systems and Structures. 2003;**14**(12): 787-793

[72] Pelrine R, Kornbluh R. Dielectric elastomers as electroactive polymers (eaps): Fundamentals. In: Capri F, editor. Electromechanically Active Polymers. Switzerland: Springer International Publishing; 2016; p. 671

[73] Zhou J, Jiang L, Khayat RE. Investigation on the performance of a viscoelastic dielectric elastomer membrane generator. Soft Matter. 2015;**11**(15):2983-2992

[74] Rudykh S, Bhattacharya K, et al. Multiscale instabilities in soft heterogeneous dielectric elastomers. In: Proceedings of the Royal Society of London A. 2014 Feb 8;**470**(2162)

[75] Zhang R, Huang X, Li T, Iravani P, Keogh P. Novel arrangements for high performance and durable dielectric elastomer actuation. Actuators. 2016;**5**(3)

[76] McKay TG, O'Brien B, Calius E, Anderson I. Self-priming dielectric elastomer generators. Smart Materials and Structures. 2010;**19**(5)

[77] Illenberger P, Takagi K, Kojima H, Madawala UK, Anderson IA. A mathematical model for self-priming circuits: Getting the most from a dielectric elastomer generator. IEEE Transactions on Power Electronics. 2017;**32**(9):6904-6912

[78] Panigrahi R, Mishra SK. An electrical model of a dielectric elastomer generator. IEEE Transactions on Power Electronics. 2018 Apr;**33**(4):2792-2797

[79] Jean-Mistral C, Porter T, Vu-Cong T, Chesne S, Sylvestre A. Modelling of soft generator combining electret and dielectric elastomer. In: 2014 IEEE/ASME International Conference on Advanced Intelligent Mechatronics (AIM). IEEE; 2014. pp. 1430-1435

[80] Huang J, Shian S, Suo Z, Clarke DR. Dielectric elastomer generator with equi-biaxial mechanical loading for energy harvesting. In: Electroactive Polymer Actuators and Devices (EAPAD). International Society for Optics and Photonics. 2013 Apr 9;**8687**: 86870Q

[81] McKay TG, Rosset S, Anderson IA, Shea H. An electroactive polymer energy harvester for wireless sensor networks. In: Journal of Physics: Conference Series. Vol. 476, No. 1. IOP Publishing; 2013

[82] Lai H, Tan CA, Xu Y. Dielectric elastomer energy harvesting and its application to human walking. In: ASME 2011 International Mechanical Engineering Congress and Exposition. American Society of Mechanical Engineers. 2011 Jan 1; pp. 601-607

[83] Moretti G, Fontana M, Vertechy R. Model-based design and optimization of a dielectric elastomer power take-off for oscillating wave surge energy converters. Meccanica. 2015; **50**(11):2797-2813

[84] Moretti G, Righi M, Vertechy R, Fontana M. Fabrication and test of an inflated circular diaphragm dielectric elastomer generator based on pdms rubber composite. Polymer. 2017;**9**(7):283

[85] Vertechy R, Fontana M, Papini GPR, Forehand D. In-tank tests of a dielectric elastomer generator for wave energy harvesting. In: Electroactive Polymer Actuators and Devices (EAPAD) 2014. Vol. 9056. International Society for Optics and Photonics; 2014

[86] Binh PC, Ahn KK. Performance optimization of dielectric electro active polymers in wave energy converter application. International Journal of Precision Engineering and Manufacturing. 2016;**17**(9):1175-1185

[87] Yurchenko D, Val DV, Lai ZH, Gu G, Thomson G. Energy harvesting from a de-based dynamic vibro-impact system. Smart Materials and Structures. 2017;**26**(10)

[88] Yurchenko D, Lai ZH, Thomson G, Val DV, Bobryk RV. Parametric study of a novel vibro-impact energy harvesting system with dielectric elastomer. Applied Energy. 2017;**208**:456-470

Chapter 5

Piezoelectric Energy Harvesting Based on Bi-Stable Composite Laminate

Fuhong Dai and Diankun Pan

Additional information is available at the end of the chapter

http://dx.doi.org/10.5772/intechopen.76193

Abstract

Energy harvesting employing nonlinear systems offers considerable advantages comparing to linear systems in the field of broadband energy harvesting. Bi-stable piezoelectric energy harvesters have been proved to be a good candidate for broadband frequency harvesting due to their highly geometrically nonlinear response during vibrations. These bi-stable energy harvesters consist of bi-stable structure and piezoelectric transducers. The nonlinear response depends on the host bi-stable structure. A possible category of bi-stable structures is bi-stable composite laminate, which has two stable equilibrium states resulting from the mismatch in thermal expansion coefficients between plies. It has received considerable interest in deformable structure since the "snap-through" between two stable states results in a significant deformation without continuous energy supply. Combining piezoelectric transducer and the bi-stable composite laminate is a feasible method to obtain bi-stable energy harvester. Piezoelectric energy harvesters based on bi-stable composite laminate have been shown to exhibit high levels of power output over a wide range of frequencies. This chapter aims to summarize and review the various approaches in piezoelectric energy harvesting based on bi-stable composite laminates.

Keywords: energy harvesting, bi-stable, piezoelectric, laminate, nonlinear

1. Introduction

With the advent of low power, wireless and autonomous sensors, energy harvesting which is considered as a potential way to replace batteries and realize self-powering becomes a highly active research area [1, 2]. Piezoelectric materials can be embedded in the host structure to convert the strain energy of the host structure into electrical energy through direct piezoelectric effect, so piezoelectric energy harvesting technique becomes one of the primary methods to

harvest vibration energy. Compared with electromagnetic and electrostatic methods, the main advantages of piezoelectric energy harvesting technique are the larger power density and higher flexibility of being integrated into one system [3]. Because of the simple structure and ease of producing relatively high average strain for a given force input, the typical piezoelectric energy harvester consisting of a cantilever beam with piezoelectric elements attached near its clamped end is widely analyzed and designed. The cantilever type harvester operates on the fundamental principle of linear resonance. It means that the maximum energy transduction from the vibration source to harvester can be achieved by tuning the host beam's natural frequencies to be equal or very close to the excitation frequency. Therefore, the frequency bandwidth of linear harvester is usually limited to a specific range and the power output of linear harvesters would be reduced drastically when the frequency of vibration source deviates slightly from the resonant frequency of the harvesters. However, the energy of ambient vibrations is distributed over a broad spectrum of frequencies or the dominant frequencies drift with time in many applications. Several solutions were presented to solve this problem. Tuning mechanism is a potential method for broadband energy harvesting, which uses passive or active means to vary the fundamental frequency of the harvester to match the dominant frequency of the vibration source [4, 5]. However, it is not very efficient when the frequency of vibration is random or varies rapidly, and tuning mechanism requires external power or complicated design. Nonlinear harvesters have been proposed for broadband energy harvesting benefiting from the ability of nonlinearities to extend the coupling between the excitation and a harmonic oscillator to a broader range of frequency.

The nonlinear harvester with a bi-stable potential is proved to be a good candidate for broadband energy harvesting. It has been shown that when carefully designed, bi-stable energy harvesters can provide significant power levels over a wide range of frequencies under steady-state harmonic excitation. Bi-stable systems have two stable equilibrium positions between which they may snap-through under a certain level of excitation. Generally, the bi-stable harvester can be achieved by applying magnetic force [6] or axial load [7] to buckle the piezoelectric beam. The potential energy of bi-stability can be tuned by the magnitude of magnetic force or axial load. However, such a magnetic bi-stable system would require an obtrusive arrangement of external magnets and could generate unwanted electromagnetic fields [8]. An alternative bi-stable composite laminate has been developed for broadband energy harvesting. This review aims to review the piezoelectric energy harvesting technologies based on two different bi-stable composite laminates and find out the potential benefits and defects from the existing energy harvesting techniques using bi-stable composite laminates.

2. Bi-stable composite laminate

The bi-stable asymmetric composite laminate was reported by Hyer firstly in 1981 [9, 10]. He found that thin asymmetric laminate may have two stable cylindrical shapes, which are attributed to the thermal stresses due to the difference of thermal expansion of the laminate. Due to the limitation of the classical lamination theory which predicts that all asymmetric laminates have a curved saddled shape where the two curvatures are always of opposite sign,

Hyer developed a theory to explain the characteristics of the curved shapes of thin asymmetric laminates. This theory introduced the von-Karman geometric nonlinearities within the classical lamination theory to capture room-temperature shapes, and Rayleigh-Ritz method based on the concept of minimum total potential energy was used to obtain the curved shapes. The prediction of room-temperature shapes of asymmetric laminate becomes one of major research directions in next decades. The analytical model can be improved by more reasonable hypothesis for displacements and strains, so accuracy and efficiency can be achieved in a good balance [11, 12]. Additionally, the development of finite element method makes it possible to predict room-temperature shapes of asymmetric laminate with more complex geometry conditions [13, 14]. Another interesting feature of bi-stable laminate is snap-through behavior, which was studied extensively [15, 16]. Recently, the potential applications of the bi-stable laminate have received considerable attention. One of potential applications is morphing structure. The main advantage of the bi-stable laminate as a morphing structure is that it has two stable positions where the structure can maintain without demanding an external power. Moreover, it only needs a very small energy input to trigger it snap from one stable position to the other with a relatively large deflection. Many researchers have studied the feasibility of bi-stable laminate as morphing structure from different perspectives, such as actuation method [17] and dynamics [18]. Another potential application of the bi-stable laminate is energy harvesting. Bi-stable laminate can provide large structural deformation resulting from the snap-through behavior. The piezoelectric elements will obtain large strain and produce high electric power if the piezoelectric elements are attached to the surface of bi-stable laminate. Compared to using magnetic mechanism, this bi-stable energy harvester has four main advantages: (1) the arrangement can be designed to occupy a smaller space; (2) there are no magnetic fields; (3) the laminate can be easily combined with piezoelectric materials; (4) there is potential to control over harvester response through adjusting lay-up and geometry [19].

According to the lay-up and stable shapes, bi-stable harvester based on composite laminate can be classified into two categories. The first one is the asymmetric laminate-based harvester. This asymmetric bi-stable laminate is made from a carbon fiber reinforced polymer (CFRP) with a $[0/90]_T$ layup, as shown in **Figure 1(a)**. For asymmetric laminate, the residual thermal stress causing curved deformation results from the differences in the thermal expansion coefficient between the carbon fiber and epoxy matrix during cooling process from an elevated cure temperature to room temperature. When the ratio of edge length to thickness increases to

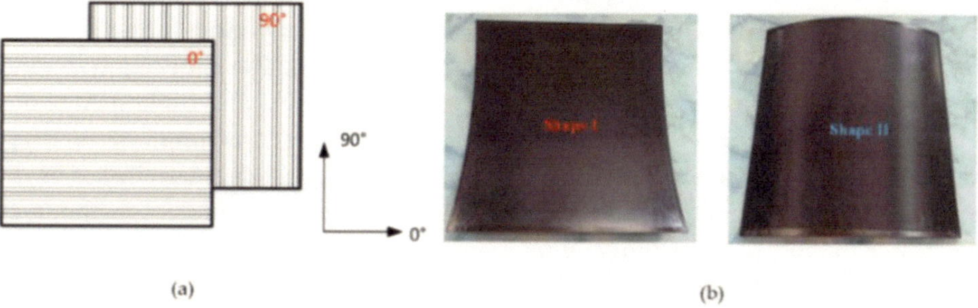

Figure 1. (a) Stacking sequence $[0/90]_T$ of the asymmetric laminate; (b) two stable shapes of the asymmetric laminate.

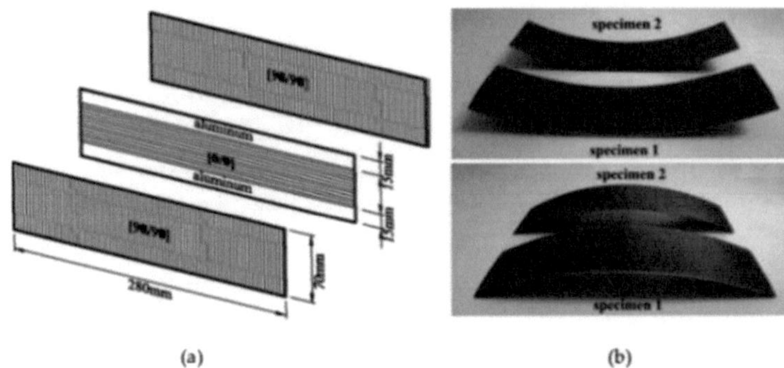

Figure 2. (a) Symmetric stacking sequence $[90_2/Al/90_2]_T \cup [90_2/0_2/90_2]_T \cup [90_2/Al/90_2]_T$ for a bi-stable hybrid symmetric laminate; (b) stable shapes of two bi-stable hybrid symmetric laminate [20].

a specific value, asymmetric laminates has two approximately cylindrical stable shapes, as shown in **Figure 1(b)**. The curvature direction of two shapes is orthogonal to each other, but the magnitude of curvature is equal to each other.

The other category is the hybrid symmetric laminate-based harvester. Li et al. [20] presented a bi-stable hybrid symmetric laminate (BHSL) by combining of aluminum plies and CFRP plies. The bi-stability is induced by the difference in thermal expansion coefficients between aluminum and CFRP. Typically, the stacking sequence of BHSL is shown in **Figure 2(a)**. There are two types of region, which are the hybrid region with $[90_2/Al/90_2]_T$ and composite region with $[90_2/0_2/90_2]_T$. The stacking sequence of two regions is symmetric. This hybrid symmetric laminate exhibits two stable cylindrical shapes with same longitudinal curvature in the opposite direction, as shown in **Figure 2(b)**.

3. Piezoelectric energy harvesting based on bi-stable laminate

3.1. Asymmetric laminate-based harvester

Asymmetric laminate has been investigated for several decades. The prediction of room-temperature shape and bifurcation phenomena with different parameters (layup, geometry, thickness and so on) can be achieved a reasonable level of accuracy with reliable theoretical model and finite element method. The new challenge for asymmetric laminate is to find a suitable application scenario according to its particular features. The application of morphing structure was firstly presented, for example, morphing airfoil [21] and trailing edge box [22]. Though asymmetric laminate is a good candidate for morphing structure, how to trigger the snap-through behavior becomes a problem. To address this problem, many types of research have studied the snap-through motion of asymmetric laminate with different methods, such as piezoelectric actuation [23], memory alloy actuation [24] and heating actuation [25]. The method adopted in piezoelectric actuation is that applying voltage on piezoelectric patches attached to the surface of laminate enables piezoelectric patches to deform. Thus, asymmetric laminate can snap once the applying voltage is high enough. Piezoelectric actuation utilizes the direct piezoelectric effect. On the contrary, the voltage will be generated by piezoelectric patches resulting from inverse piezoelectric effect when asymmetric laminate deforms under external excitation. Therefore, the application of asymmetric laminate extends to piezoelectric energy harvesting.

Arrieta et al. [26] firstly presented a piezoelectric nonlinear broadband energy harvester based on asymmetric laminate in 2010. Only the experimental results were reported in this earlier work. This square 200 × 200 mm asymmetric laminate with $[90_2/0_2]_T$ layup was mounted from its center to an electromechanical shaker, and four PZT-5A flexible piezoelectric patches were bonded to the surface of the laminate, as shown in **Figure 3**.

Five types of nonlinear responses were observed in the experiments. As shown in **Figure 4(a)**, these five types of responses are linear oscillation, large amplitude limit cycle oscillation (LCO), chaotic oscillation, intermittency oscillation and 1/2 subharmonic oscillation. Large amplitude LCO, chaotic oscillation, and intermittency oscillation involve snap-through behavior between two stable shapes, so the voltage output corresponding to these three types of

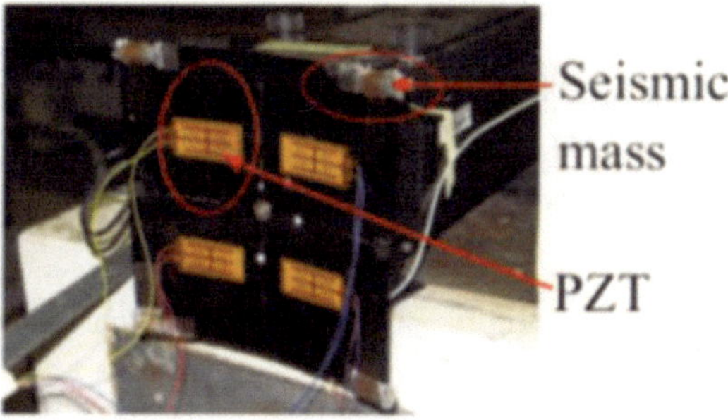

Figure 3. Piezoelectric energy harvester based on asymmetric laminate [26].

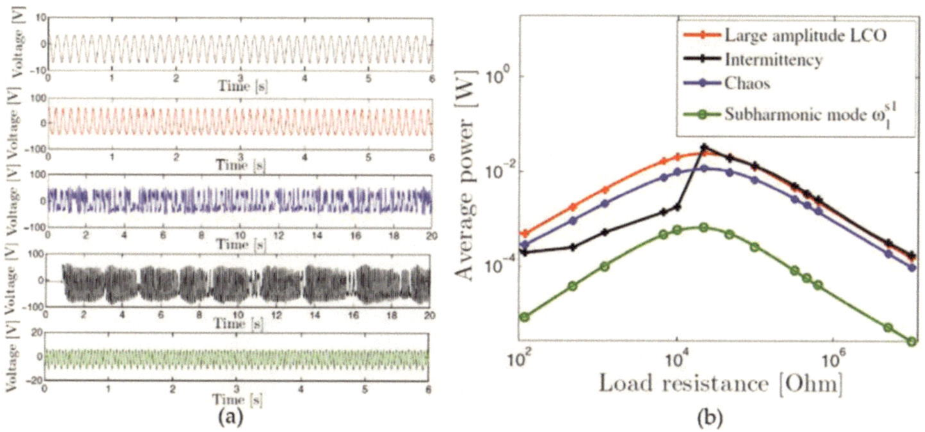

Figure 4. (a) Different responses for a forcing level 2 g: Linear response at 7.5 Hz; large amplitude LCO at 8.6 Hz; chaotic oscillations at 12.5 Hz; intermittency oscillations at 9.8 Hz; 1/2 subharmonic oscillations at 20.2 Hz. (b) Experimental average power vs. load resistance for different responses [26].

responses are apparently higher than remaining two responses. Accordingly, the output powers of these three types are also higher, as shown in **Figure 4(b)**. The intermittency and large amplitude LCO can give 34 and 27 mW respectively under a forcing acceleration level of 2.0 g. For chaotic oscillation, the maximum average power was found to be 9 mW. The experimental results in this work illustrate that asymmetric bi-stable laminate has rich dynamics for nonlinear energy harvesting and it is a potential candidate for broadband energy harvesting through combining the ranges where large amplitude LCO, chaotic oscillation, and intermittency oscillation occur.

Betts et al. [8, 27, 28] published several works to optimize configurations of asymmetric laminate for improving electrical power generation in 2012. This piezo-laminate was considered in the actuation arrangement, as shown in **Figure 5**. The actuation force applied on the center of the laminate surface while displacements of all four corners were constrained in the z-direction. The optimization was based on the static states of the system, so the electrical energy output depending on the snap-through between two stable shapes. The M8557-P1 MFC was employed as a piezoelectric material. Maximization of electrical energy generated in two sets of four piezoelectric layers was as optimizable objective under several constraint conditions, such as laminate surface area, limitation of piezoelectric strain and bistability. Four variables were adopted which are ply orientation, single ply thickness, aspect ratio and piezoelectric surface area. It was found that square cross-ply laminates $[0^P/0/90/90^P]_T$ offered the most extensive energy outputs while the laminate curvatures were maximized and aligned with the piezoelectric polarization axis.

The results of optimization are meaningful for the design of bi-stable energy harvester based on asymmetric laminate, but the optimization was carried out from a static perspective. The energy harvesting from vibration is a dynamic problem, so the optimization results are not very suitable for dynamic situations. Therefore, Betts et al. [29] had done some preliminary works about the dynamic transition between stable states as the piezoelectric laminate was exposed to an oscillating mechanical force in the same year. It was found that thicker laminates produced higher levels of energy when snap-through is fully induced. Furthermore, the dynamic loading causes a higher level of strains of piezoelectric transducers, which may cause its failure.

Next year, Arrieta et al. [30, 31] presented a new concept for broadband energy harvesting. They changed the boundary condition from center fixation to cantilever for asymmetric

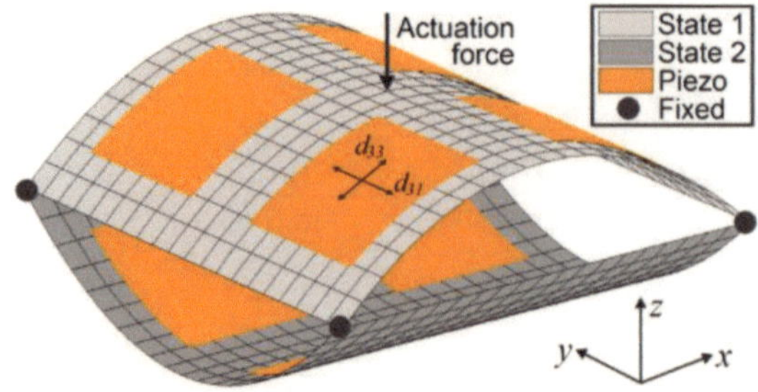

Figure 5. Actuation arrangement for a $[0^P/0/90/90/90^P]_T$ laminate with 40% piezoelectric coverage [27].

laminate. A symmetric-asymmetric layup was designed to realize cantilevered boundary condition, as shown in **Figure 6(a)**. The symmetric layup was for clamping, and the asymmetric layup was for bi-stable laminate. Two flexible piezoelectric transducers (Piezo QP16n) were bonded on the surface of the asymmetric laminate and were closer to the clamped root. The two stable shapes of this cantilevered bi-stable laminate are shown in **Figure 6(b)**. There are three significant advantages to this design. Firstly, this type of arrangement allows to exploit high strains developed close to clamped root by the piezoelectric transducers; Secondly, the cantilever configuration results in large displacements given the considerable distance from the root to the tip; Thirdly, the cantilevered configuration allows for more natural integration with the host structure.

Two specimens with different length were fabricated and tested respectively. The nonlinear behaviors of two specimens were studied by frequency sweep for a base acceleration, as shown in **Figure 7** where blue dots and red crosses show frequency sweep with the initial condition on state 1 and 2 respectively, and arrows show regions of cross-well dynamics. Due to the cantilevered boundary condition, the two stable shapes are entirely asymmetric which leads to the different nonlinear responses depending on the initial state. Specimen B with longer length has a broader range of cross-well oscillation than specimen A. The cantilevered bi-stable laminate can be triggered to snap at a low level of base excitation. Additionally, the obtained power of this design ranges from 35 to 55 mW when Synchronized Switching Harvesting on Inductor (SSHI) circuit is adopted.

In the same year, Betts et al. [32] extended their work to experimental investigation of dynamic response and power generation characteristics of piezoelectric energy harvester based on a square $[0/90]_T$ laminate with the size of 190 × 190 mm. A single piezoelectric Marco Fiber Composite (MFC) layer (M8585-P2, 85 × 85 mm) was attached to the laminate surface, as shown in **Figure 8(a)**. Additional masses were attached to the four corners of the laminate to increase the achievable curvatures and help snap-through during oscillation. The whole device was mounted to the shaker from its center, as shown in **Figure 8(b)**. The open-circuit voltage

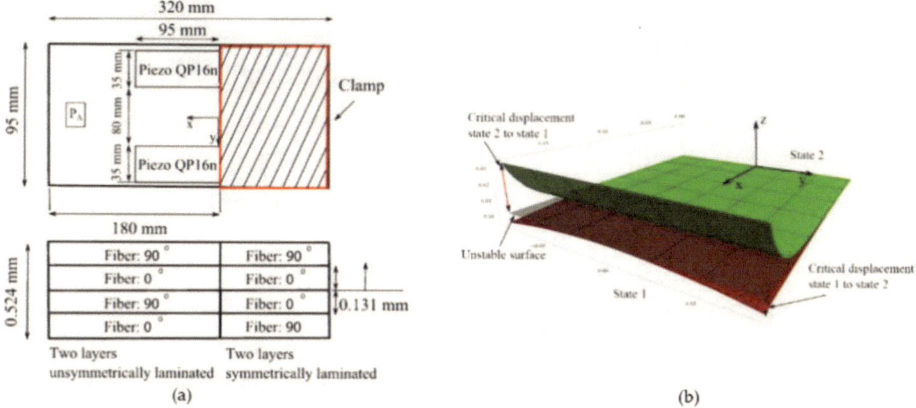

Figure 6. (a) Symmetric-asymmetric lay-up enabling cantilever configuration and positioning of piezoelectric transducers [30]. (b) Stable shapes and critical displacement for a cantilevered bi-stable laminate [31].

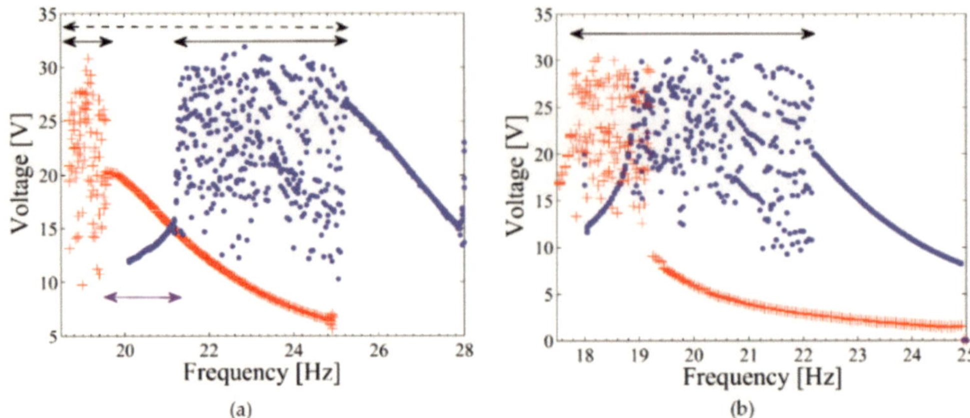

Figure 7. (a) Frequency-voltage response against forcing frequency of specimen A (L_x = 168 mm). Base acceleration level of 0.33 g, and a load resistance of 66 kΩ. (b) the frequency-voltage response against forcing frequency of specimen B (L_x = 178 mm). Base acceleration level of 0.25 g, and a load resistance of 66 kΩ [30].

Figure 8. (a) Stable shapes of a $[0/90]_T$ piezoelectric laminate; (b) experimental setup showing mechanical shaker attachment [32].

was measured, and the three-dimensional displacements were captured by a Digital Image Correlation (DIC) system. They had the similar experimental results as what was found by Arrieta et al. [26]. The experimental results have revealed that the modes of oscillation are sensitive to the frequency and amplitude of the external vibration, as shown in **Figure 9**. There are three oscillation patterns: (1) small-amplitude oscillation without snap-through; (2) uniform and nonuniform intermittent snap-through; (3) repeatable snap-through. The largest power output of 3.2 mW was found when snap-through occurred.

In 2014, Betts et al. [33] continued their work and presented an analytical model and experimental characterization of a piezoelectric bi-stable laminate of 200 × 200 × 0.5 mm with $[0/90]_T$ stacking sequence. As before, a single flexible Macro Fiber Composite (MFC) was employed as a piezoelectric transducer and attached to one surface of the laminate. The analytical model was an extension of the model presented in Ref. [34]. This analytical model can capture the mix of nonlinear modes in the response of this bi-stable piezoelectric energy harvester subjected to mechanical vibrations. As before, these modes are continuous snap-through, intermittent snap-through (both periodic and chaotic), and small amplitude oscillations which were validated by experiments.

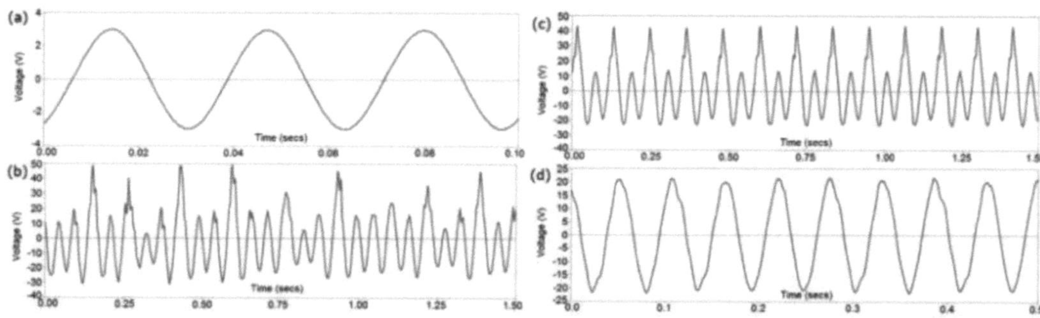

Figure 9. Different voltage outputs of bi-stable piezoelectric laminate: (a) low amplitude oscillations; (b) nonuniform behavior and chaotic snap-through; (c) intermittent snap-through; (d) repeated snap-through [32].

A map of these modes while varying the acceleration (g level) and frequency of excitation was obtained based on experimental results, as shown in **Figure 10(a)**. The results show that the desired continuous snap-through mode for energy harvesting requires high acceleration level, and intermittency snap-through modes can cover a broader range of frequency. The average powers with different frequencies and accelerations were measured and compared with analytical results. The analytical results are higher than experimental results, and analytical and experimental average powers for 10 g excitation are shown in **Figure 10(b)**. This design can cover the bandwidth of 19.4 Hz involving snap-through behavior at the acceleration of 10 g, and the peak power is as high as 244 mW. Though the power output and bandwidth are so excellent, the acceleration level demanded is much higher than the work of Arrieta et al. [30].

In 2015, Syta et al. [35] employed the same design as Betts et al. [33]. The difference is that "0–1 test" was introduced for the experimental works to identify the chaotic dynamics

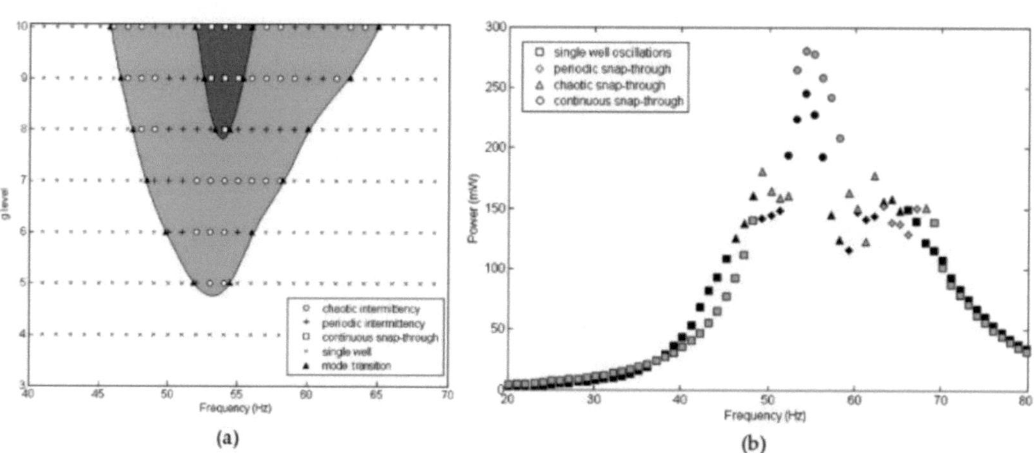

Figure 10. (a) Experimentally observed mode types associated with all combinations of drive frequency (20–80 Hz) and acceleration g-level (3–10 g); (b) average power outputs and associated modes for 10 g excitations. Experimental (black symbols) and modeling results (gray symbols) [33].

of this bi-stable energy harvester. Next year, Syta et al. [36] examined the modal responses of this bi-stable electro-mechanical energy harvester with same design by Fourier spectrum and Recurrence Quantification Analysis (RQA). RQA was used to identify the periodic and chaotic responses from reasonably short time series.

To overcome the high excitation level demand of large amplitude LCO at high frequencies, in 2015, Li et al. [37] exploited the nonlinear oscillations around the second vibration mode of a rectangular piezoelectric bi-stable laminate (RPBL) for broadband vibration energy harvesting at relatively higher frequencies with relatively lower excitation acceleration. This harvester consisted of a 150 × 50 × 0.42 mm, $[0_2/90_2]_T$ laminate and a 15 × 15 × 0.2 mm piece of PZT-5H and copper electrode, as shown in **Figure 11(a)**. Through finite element analysis, the frequency of second vibration mode of state A with mass is 66.8 Hz, and the second vibration mode shape is shown in **Figure 11(b)**. The experimental results show that the lowest excitation acceleration needed to trigger the LCO is 2.66 g at 59 Hz. Two optimized RPBLs were found by finite element analysis, which illustrates that the frequency bandwidth of LCO can be broadened by decreasing the deformation needed to trigger the local snap-through. Although this design lowers the excitation, the power output of 0.98 mW is relatively low.

In 2014, Harris et al. [38] manufactured two piezoelectric energy harvesters and compared their performance by experiments. One of the two energy harvesters is linear with an asymmetric layup, as shown in **Figure 12(a)**. The other one is bi-stable with asymmetric layup shown in **Figure 12(b)** (,) and the two stable states are shown in **Figure 12**. The results showed that the bi-stable harvester had higher power output over a broader range of frequencies at low frequency and low excitation and the linear one had the potential to produce a higher peak power but at a narrow bandwidth. In 2016, Harris et al. [19] continued their work and investigated the dynamics of this bi-stable energy harvester by multiscale entropy and "0–1" test. As before, these oscillation modes including single-well oscillation, periodic and chaotic intermittent snap-through and continuous periodic snap-through were captured in experiments. The multiscale entropy and "0–1" test can be helpful in the response characterization. One benefit from this analysis method is that the continuous plate system may be characterized by a single variable (voltage and displacement).

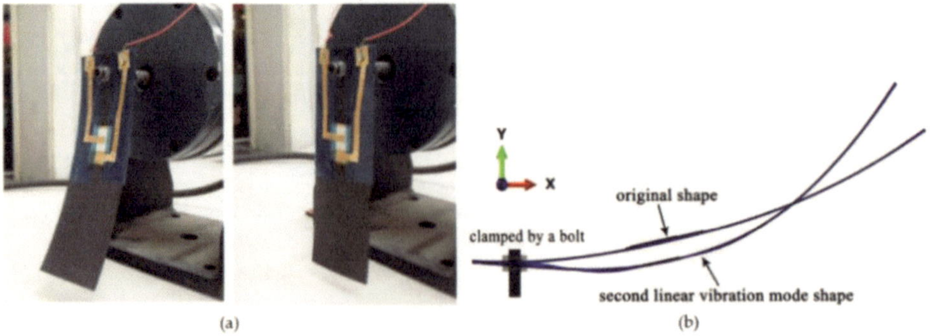

Figure 11. (a) Two stable states of RPBL; (b) the second vibration mode of stable state A (with proof mass) [37].

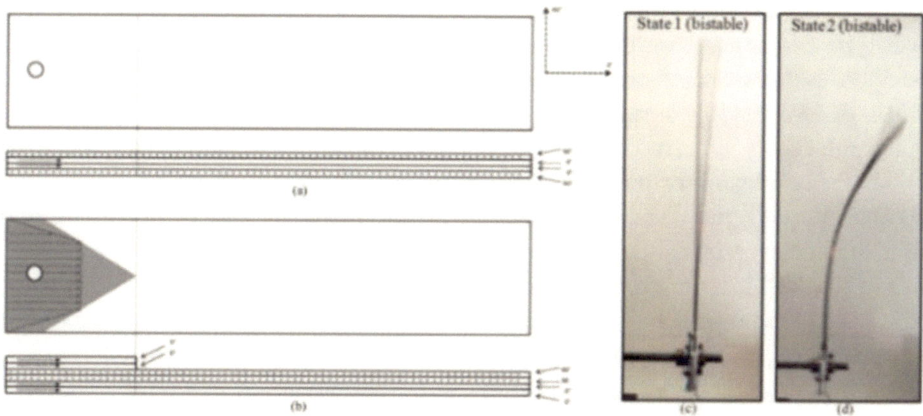

Figure 12. Laminate lay-ups: (a) linear and symmetric $[90/0/0/90]_T$; and (b) bi-stable and asymmetric $[0/0/90/90]_T$. The cantilevers were clamped at the left-hand side. Bi-stable states: (c) state I; and (d) state II [38].

In the same year, Harris et al. [47] added magnets in the bi-stable system to lower the level of excitation that triggers the snap-through for this bi-stable cantilever from one stable state to another, as shown in **Figure 13**. The system performance can be adjusted by varying the separation between the magnets. The scenario without magnets was taken as a control, and the different separations were measured. The results showed that this approach could adjust the fundamental frequency of the harvester and the magnets benefited the increase of the bandwidth with lower peak power at low acceleration levels and benefited the increase of peak power with narrower bandwidth at higher excitation levels. Additionally, a single-degree-of-freedom (SDOF) model was established based on the experimental load-deflection characteristic.

3.2. Hybrid symmetric laminate-based harvester

Besides traditional asymmetric bi-stable laminate, bi-stable laminates with novel layup supply new potential ways to design energy harvester. In 2015, Pan et al. [40] presented a bi-stable piezoelectric energy harvester (BPEH) based on bi-stable hybrid symmetric laminate (BHSL). The most apparent difference between asymmetric bi-stable laminate and BHSL is the stable shape. BHSL has two double-curved shapes, which have identical curvatures with opposite signs, as shown in **Figure 2**. It has better designability compared with the traditional

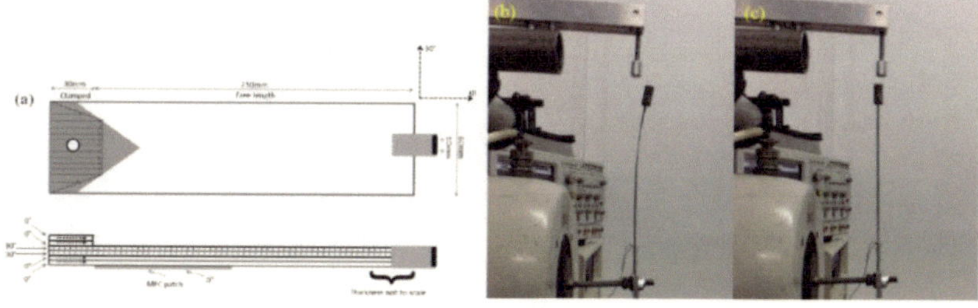

Figure 13. (a) Detail of the ply orientations, magnet location, and MFC location; (b) stable state I and (c) stable state II [39].

asymmetric laminate due to the variation of the position of the metallic layer. Moreover, the cantilever-type boundary condition can be realized benefiting from its unique symmetric stable shapes. In this work, to obtain more deformation, 20 pieces of PZT-5H were bonded to the middle of two BHSL's surfaces where the curvatures distribute uniformly. Due to the electrical conductivity of carbon fiber, the laminate can be as an electrode, and a parallel connection was employed for piezoelectric transducers. Two types of stacking sequences and two types of piezoelectric transducer shapes with the identical area were chosen, and four types of bi-stable harvesters and one linear harvester (LPEH) as control sample were designed, as shown in **Figure 14**. Through finite element analysis, it was found that stress mechanism of PZTs consists of bending stress and residual thermal stress and the stable shapes cannot reflect the deformation of the PZT. A rough power measurement, which is handshaking to actuate continuous snap-through behavior, was employed. The maximum power of 37.06 mW with the optimal resistance of 40 kΩ was obtained at 5 Hz from the BPEH with highest open-circuit voltage. Additionally, through the comparison between BPEHs and LPEH, it was shown that the BPEH could take advantages of each piece of PZT. It means the PZTs on BHSL can obtain uniform deformations during the snap-through process, which leads that BPEH can output much higher power.

Through previous preliminary investigation, BPEH shows good potential for energy harvester. In 2017, Pan et al. [41] continued their work and investigated the dynamics of this bi-stable energy harvester. The bi-stable laminate was redesigned with a smaller size of 100×40 mm and 8 pieces of PZT-5H with 1.0×1.0 mm were bonded in the middle of laminate surface like before. Three types of linear harvesters with two different layups and two types of PZT positions were designed as control samples, as shown in **Figure 15**.

The forward sweeps and reverse sweeps at five acceleration levels were carried out for BPEH, as shown in **Figure 16**. Unlike responses of three linear harvesters, which only has have one sharp peak, responses of BPEH exhibited nonlinear characteristics. BPEH had a softening response under low level of excitation, and it could switch to hardening response when the excitation increased to a certain extent. Two oscillation modes were observed in experiments, which were single-well oscillation and cross-well oscillation. BPEH only has single-well oscillation mode under softening response at relatively low excitation level. When the response switch to hardening, BPEH can repeatedly travel between its two

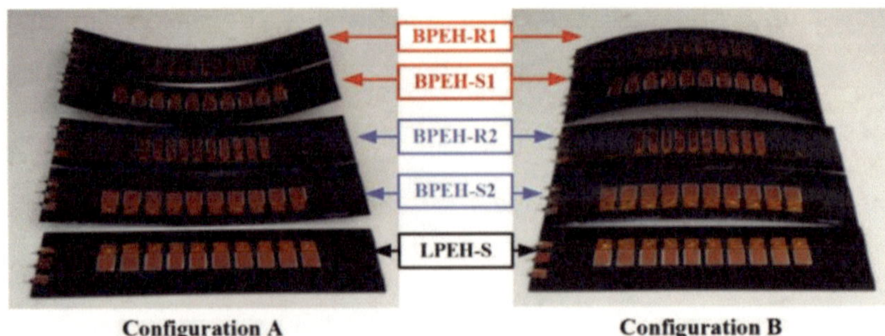

Figure 14. Stable shapes of the BPEHs and LEPH samples [40].

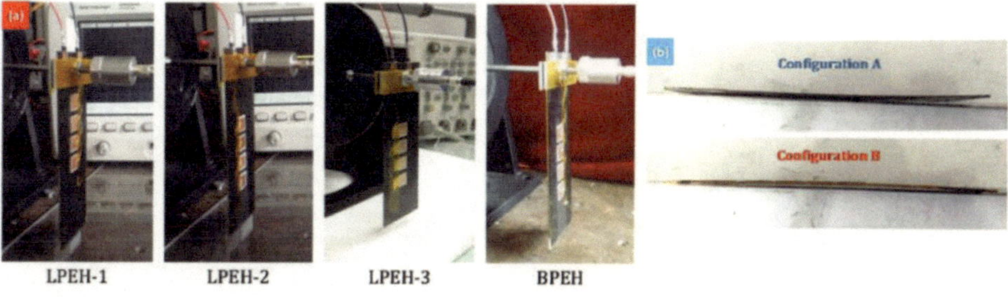

Figure 15. (a) Experimental setup showing mechanical shaker attachment of three types of LPEH and BPEH; (b) two stable configurations of BPEH [41].

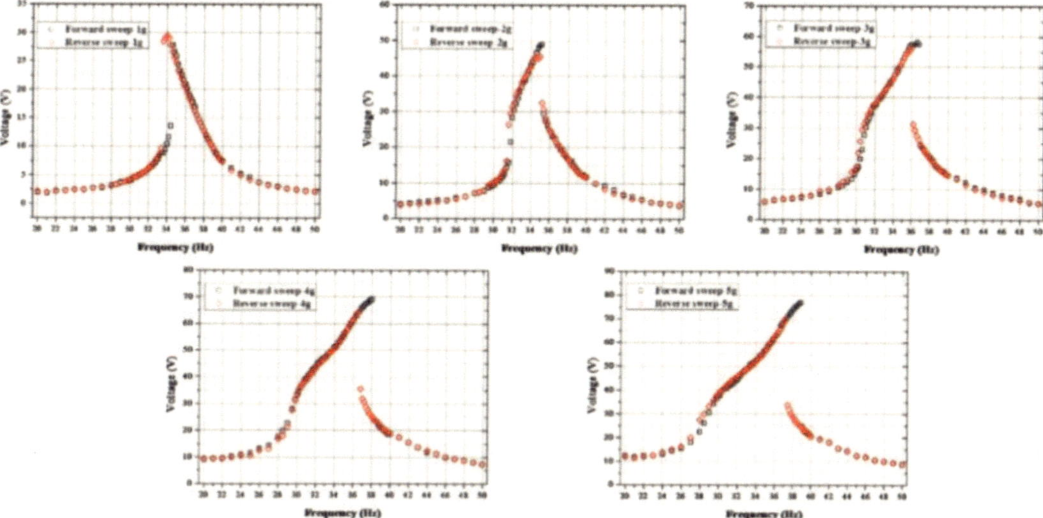

Figure 16. Experimental frequency-voltage responses of BPEH for acceleration ranging from 1 to 5 g [41].

potential wells, namely be under cross-well oscillation mode. When BPEH is under cross-well oscillation mode, the output voltage is much higher than that of single-well oscillation mode. The responses also are affected by sweep direction. The frequency range of high-voltage branch in the forward sweep is wider than that in the reverse sweep when BPEH exhibits hardening-type nonlinearity. Compared to LPEHs, BPEH has higher output voltage in a wider frequency range under the same excitation level. For output power, BPEH has a more remarkable performance than LPEHs. The maximum output power of BPEH subjected to 5 g acceleration at 36 Hz is 5.7 times more than that of LPEH-1. The reasons for this can be attributed to two aspects of higher voltage and more uniform strains in piezoelectric elements. The results of BPEH average power showed that the hardening responses help BPEH extend the high-output bandwidth and the output power associated with the excitation frequency and acceleration. All the experimental results demonstrated that BPEH has the potential to harvest vibration energy under broadband excitations.

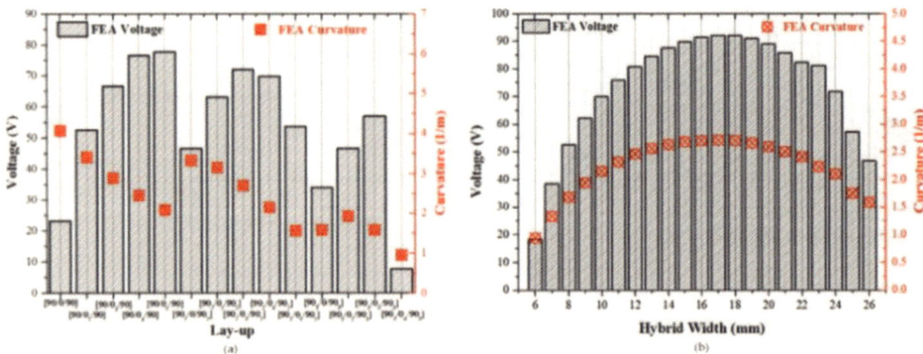

Figure 17. Initial voltages and longitudinal curvatures with (a) different lay-up and (b) hybrid width [42].

In the same year, Pan et al. [42] analyzed the influence of layup design on the performance of this bi-stable energy harvester. The initial voltage induced by stable configuration and longitudinal curvature with different lay-up and hybrid width was calculated and analyzed by a static finite element analysis. The results showed that the lay-up could vary the initial voltage and longitudinal in opposite directions, and hybrid width can adjust these two variables in the same direction, as shown in **Figure 17**. Through finite element analysis, it was found that the initial voltage of BPEH depends on strain variations in the two directions. Three types of BPEHs were manufactured to verify the analytical results. In the experiments, three types of oscillation modes were observed which are continuous cross-well vibration, single-well vibration, and intermittent cross-well vibration. The inherent characteristics of BPEH determine the frequency of characterized voltage, and the stable configuration affects the vibration mode. The BPEH with the lowest curvature can occur continuous snap-through vibration, but the BPEH with highest initial voltage only can occur single-well vibration. The layup and hybrid width can affect inherent characteristics and configurations at some time. The combination of lower frequency and lower longitudinal curvature is easier to obtain desired continuous cross-well vibration for energy harvesting.

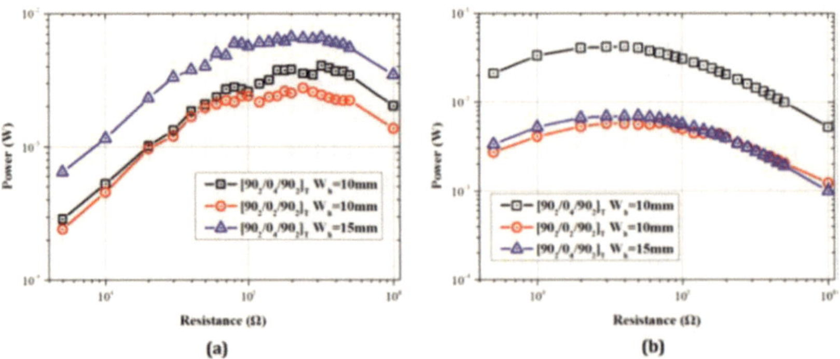

Figure 18. Experimental output power dependence of resistance for BPEHs (a) hand-driven method; (b) shaker-driven method [42].

	Laminate type and size	Piezoelectric transducer type and size	Boundary condition	Vibration mode	Excitation level and frequency	Maximum power	Power density	Bandwidth
Arrieta et al. [26]	Asymmetric 20×20 cm^2	Quick pack QP 16n $4.6 \times 2.06 \times 0.0254$ cm$^3 \times 4$	Center fixation	Large amplitude LCO	2 g 8.6 Hz	27 mW[a]	28.0 mW/cm^{-3}	2.2 Hz (8–10.2 Hz)
Betts et al. [32]	Asymmetric 19×19 cm^2	MFC M8585-P2 $8.5 \times 8.5 \times 0.03$ cm$^3 \times 1$	Center fixation	Repeatable snap-through	2.3 g 18 Hz	3.2 mW[a]	2.2 mW/cm^{-3}	—
Arrieta et al. [30]	Asymmetric 18×9.5 cm^2	Quick pack QP 16n $4.6 \times 2.06 \times 0.0254$ cm$^3 \times 2$	Cantilever	Chaotic oscillation	0.25 g 20.5 Hz	5 mW[a] 55 mW[c]	10.4 mW/cm^{-3} 114.25 mW/cm^{-3}	4.5 Hz
Betts et al. [33]	Asymmetric 20×20 cm^2	MFC M8585-P2 $8.5 \times 8.5 \times 0.03$ cm$^3 \times 1$	Center fixation	Continuous snap-through	10 g 54 Hz	244 mW[a]	112.57 mW/cm^{-3}	22 Hz
Li et al. [37]	Asymmetric 15×5 cm^2	PZT-5H $1.5 \times 1.5 \times 0.02$ cm$^3 \times 1$	Clamped at the end by a bolt	LCO (2nd Mode)	3.1 g 59 Hz	0.98 mW[b]	21.8 mW/cm^{-3}	10 Hz (59–69 Hz by FEA)
Pan et al. [40]	Hybrid Symmetric 30×6 cm^2	PZT-5H $2.5 \times 0.9 \times 0.02$ cm$^3 \times 20$	Hold by hand	Continuous snap-through	By hand shaking 5 Hz	37.06 mW[b]	41.18 mW/cm^{-3}	—
Harris et al. [38]	Asymmetric 25×6 cm^2	MFC M8528-P2 $10.5 \times 3.4 \times 0.03$ cm$^3 \times 1$	Cantilever	Snap-through	6 g 8.4 Hz	7.3 mW[a]	6.81 mW/cm^{-3}	8.4 Hz (FWHM)
Harris et al. [39]	Asymmetric 21×6 cm^2	MFC M8528-P2 $10.5 \times 3.4 \times 0.03$ cm$^3 \times 1$	Cantilever	Snap-through	3 g	4.51 mW[a]	4.21 mW/cm^{-3}	5.6 Hz
Pan et al. [41]	Hybrid Symmetric 10×4 cm^2	PZT-5H $1.0 \times 1.0 \times 0.02$ cm$^3 \times 8$	Cantilever	Continuous cross-well	5 g 38 Hz	36.2 mW[b]	226.25 mW/cm^{-3}	8.4 Hz (30.4–38.8 Hz)
Pan et al. [42]	Hybrid Symmetric 20×6 cm^2	PZT-5H $1.5 \times 1.5 \times 0.02$ cm$^3 \times 12$	Cantilever	Continuous cross-well	6 g 16 Hz	42.2 mW[b]	78.1 mW/cm^{-3}	4 Hz

[a]Root mean square power.
[b]the output power is measured via ac-dc circuit.
[c]the output power is measured via SSHI.

Table 1. Comparison of the reviewed piezoelectric energy harvesters based on bi-stable composite laminate.

Three types of BPEH were actuated by two methods, which are a hand-driven method and a shaker-driven method, respectively. The hand-driven method was utilized to confirm the contribution of initial voltage to output power. As shown in **Figure 18(a)**, BPEH with higher initial voltage can generate higher power by the hand-driven method as expected. However, the different results were found by the shaker-driven method, as shown in **Figure 18(b)**. The BPEH under continuous cross-well vibration outputs the highest power. However, BPEH with the highest initial voltage only can occur single-well vibration so that it has the lower power. BPEH, which is under intermittent cross-well vibration, generates the lowest maximum power.

4. Summary

This chapter reviewed the piezoelectric energy harvesting technique based on bi-stable composite laminate. As an essential branch of bi-stable energy harvesting, composite laminate has its unique advantages, such as inherent nonlinearity and combination with piezoelectric materials. The comparison of the reviewed piezoelectric energy harvester based on bi-stable composite laminate is shown in **Table 1**. Most of the designs are based on asymmetric laminate due to its more mature theory and design method. However, asymmetric laminate has observable bifurcation phenomena of size so that current designs have a relatively large size and it is hard to decrease the size. It may limit its further application. Hybrid laminate has a more flexible design and has an excellent potential to decrease the size. However, the theories including statics and dynamics of the hybrid laminate are still lacked. Also, most of the designs employ composite piezoelectric materials (MFC) as a transducer of which useful piezoelectric volume is limited. It is against to objective of decreasing the size and maintaining outputs. So far, the required excitation level of large amplitude oscillation (snap-through) is high. How to lower the required excitation with guaranteed reliability, stability and performance is a challenging task in both theory and practice. The harvesters based on bi-stable composite laminate have a relatively high output power (mW level) which can satisfy the requirement of conventional wireless sensors. It is difficult to evaluate their quality whether good or bad due to the different characterization method of power. However, portability design and low demanding are hot topics in this area of energy harvesting. Another way for harvester based on bi-stable laminate is finding suitable application situation, such as large amplitude and low-frequency vibration.

Acknowledgements

This work was supported by National Natural Science Foundation of China [grant No. 11372087], China Scholarship Council, Ph.D. Programs Foundation of Ministry of Education of China.

Author details

Fuhong Dai* and Diankun Pan

*Address all correspondence to: daifh@hit.edu.cn

National Key Laboratory of Science and Technology on Advanced Composites in Special Environment, Harbin Institute of Technology, Harbin, China

References

[1] Kim HS, Kim J-H, Kim J. A review of piezoelectric energy harvesting based on vibration. International Journal of Precision Engineering and Manufacturing. 2011;**12**(6):1129-1141

[2] Sodano HA, Inman DJ, Park G. A review of power harvesting from vibration using piezoelectric materials. Shock and Vibration Digest. 2004;**36**(3):197-206

[3] Li H, Tian C, Deng ZD. Energy harvesting from low frequency applications using piezoelectric materials. Applied Physics Reviews. 2014;**1**(4):041301

[4] Roundy S, Zhang Y. Toward self-tuning adaptive vibration-based microgenerators. Smart Materials, Nano-, and Micro-Smart Systems, SPIE. 2005. pp. 373-384

[5] Challa VR, Prasad M, Shi Y, Fisher FT. A vibration energy harvesting device with bidirectional resonance frequency tunability. Smart Materials and Structures. 2008;**17**(1):015035

[6] Erturk A, Hoffmann J, Inman D. A piezomagnetoelastic structure for broadband vibration energy harvesting. Applied Physics Letters. 2009;**94**(25):254102

[7] Masana R, Daqaq MF. Relative performance of a vibratory energy harvester in mono-and bi-stable potentials. Journal of Sound and Vibration. 2011;**330**(24):6036-6052

[8] Betts DN, Kim HA, Bowen CR. Preliminary study of optimum piezoelectric cross-ply composites for energy harvesting. Smart Materials Research. 2012;**2012**:8. Article ID 621364. DOI: 10.1155/2012/621364

[9] Hyer MW. Some observations on the cured shape of thin unsymmetric laminates. Journal of Composite Materials. 1981;**15**(2):175-194

[10] Hyer MW. The room-temperature shapes of four-layer unsymmetric cross-ply laminates. Journal of Composite Materials. 1982;**16**(4):318-340

[11] Cho M, Kim M-H, Choi HS, Chung CH, Ahn K-J, Eom YS. A study on the room-temperature curvature shapes of unsymmetric laminates including slippage effects. Journal of Composite Materials. 1998;**32**(5):460-482

[12] Jun W, Hong C. Cured shape of unsymmetric laminates with arbitrary lay-up angles. Journal of Reinforced Plastics and Composites. 1992;**11**(12):1352-1366

[13] Schlecht M, Schulte K, Hyer M. Advanced calculation of the room-temperature shapes of thin unsymmetric composite laminates. Composite Structures. 1995;**32**(1):627-633

[14] Schlecht M, Schulte K. Advanced calculation of the room-temperature shapes of unsymmetric laminates. Journal of Composite Materials. 1999;**33**(16):1472-1490

[15] M-L D, Hyer MW. The response of unsymmetric laminates to simple applied forces. Mechanics of Composite Materials and Structures: An International Journal. 1996;**3**(1):65-80

[16] Cantera M, Romera J, Adarraga I, Mujika F. Modelling and testing of the snap-through process of bi-stable cross-ply composites. Composite Structures. 2015;**120**:41-52

[17] Pirrera A, Avitabile D, Weaver P. On the thermally induced bistability of composite cylindrical shells for morphing structures. International Journal of Solids and Structures. 2012; **49**(5):685-700

[18] Arrieta AF, Neild SA, Wagg DJ. On the cross-well dynamics of a bi-stable composite plate. Journal of Sound and Vibration. 2011;**330**(14):3424-3441

[19] Harris P, Bowen C, Kim A, Litak G. Dynamics of a vibrational energy harvester with a bi-stable beam: Voltage response identification by multiscale entropy and "0–1" test. The European Physical Journal Plus. 2016;**131**(4):109

[20] Li H, Dai F, Weaver PM, Du S. Bi-stable hybrid symmetric laminates. Composite Structures. 2014;**116**:782-792

[21] Portela P, Camanho P, Weaver P, Bond I. Analysis of morphing, multi stable structures actuated by piezoelectric patches. Computers & Structures. 2008;**86**(3):347-356

[22] Diaconu CG, Weaver PM, Mattioni F. Concepts for morphing airfoil sections using bi-stable laminated composite structures. Thin-Walled Structures. 2008;**46**(6):689-701

[23] Schultz MR, Hyer MW. Snap-through of unsymmetric cross-ply laminates using piezoceramic actuators. Journal of Intelligent Material Systems and Structures. 2003;**14**(12):795-814

[24] Dano ML, Hyer MW. SMA-induced snap-through of unsymmetric fiber-reinforced composite laminates. International Journal of Solids and Structures. 2003;**40**(22):5949-5972

[25] Li H, Dai F, Du S. Numerical and experimental study on morphing bi-stable composite laminates actuated by a heating method. Composites Science and Technology. 2012;**72**(14): 1767-1773

[26] Arrieta A, Hagedorn P, Erturk A, Inman D. A piezoelectric bi-stable plate for nonlinear broadband energy harvesting. Applied Physics Letters. 2010;**97**(10):104102

[27] Betts DN, Kim HA, Bowen CR, Inman D. Optimal configurations of bi-stable piezo-composites for energy harvesting. Applied Physics Letters. 2012;**100**(11):114104

[28] Betts DN, Kim HA, Bowen CR, Inman DJ. Optimization of piezoelectric bistable composite plates for broadband vibrational energy harvesting. Proc. SPIE 8341, Active and Passive Smart Structures and Integrated Systems. 2012, 83412Q (28 March 2012); DOI: 10.1117/12.930138; https://doi.org/10.1117/12.930138

[29] Betts DN, Kim HA, Bowen CR, Inman DJ. Static and dynamic analysis of bistable piezoelectric- composite plates for energy harvesting. In: 53rd AIAA/ASME/ASCE/AHS/ASC Structures, Structural Dynamics and Materials Conference, 2012-04-23 - 2012-04-26. American Institute of Aeronautics and Astronautics (AIAA). 2012

[30] Arrieta A, Delpero T, Bergamini A, Ermanni P. Broadband vibration energy harvesting based on cantilevered piezoelectric bi-stable composites. Applied Physics Letters. 2013;**102**(17):173904

[31] Arrieta AF, Delpero T, Bergamini A, Ermanni P. A cantilevered piezoelectric bi-stable composite concept for broadband energy harvesting. Proc. SPIE 8688, Active and Passive Smart Structures and Integrated Systems 2013, 86880G (10 April 2013); DOI: 10.1117/12.2010122; https://doi.org/10.1117/12.2010122

[32] Betts DN, Bowen CR, Kim HA, Gathercole N, Clarke CT, Inman DJ. Nonlinear dynamics of a bi-stable piezoelectric-composite energy harvester for broadband application. The European Physical Journal Special Topics. 2013;**222**(7):1553-1562

[33] Betts DN, Bowen CR, Kim HA, Guyer RA, Le Bas P-Y, Inman DJ. Modelling the dynamic response of bi-stable composite plates for piezoelectric energy harvesting. 55th AIAA/ASME/ASCE/AHS/ASC Structures, Structural Dynamics, and Materials Conference, 2014.2014-01-13 - 2014-01-17, National Harbor

[34] Diaconu CG, Weaver PM, Arrieta AF. Dynamic analysis of bi-stable composite plates. Journal of Sound and Vibration. 2009;**322**(4):987-1004

[35] Syta A, Bowen C, Kim H, Rysak A, Litak G. Experimental analysis of the dynamical response of energy harvesting devices based on bi-stable laminated plates. Meccanica. 2015;**50**(8):1961-1970

[36] Syta A, Bowen CR, Kim HA, Rysak A, Litak G. Responses of bi-stable piezoelectric-composite energy harvester by means of recurrences. Mechanical Systems and Signal Processing. 2016;**76-77**:823-832

[37] Li H, Dai F, Du S. Broadband energy harvesting by exploiting nonlinear oscillations around the second vibration mode of a rectangular piezoelectric bi-stable laminate. Smart Materials and Structures. 2015;**24**(4):045024

[38] Harris P, Bowen CR, Kim HA. Manufacture and characterisation of piezoelectric broadband energy harvesters based on asymmetric bi-stable laminates. Journal of Multifunctional Composites. 2014;**2**(3):113-123

[39] Harris P, Litak G, Bowen CR, Arafa M. A composite beam with dual bistability for enhanced vibration energy harvesting. Proc. SPIE 9865, Energy Harvesting and Storage:

Materials, Devices, and Applications VII, 98650K (17 May 2016); DOI: 10.1117/12.2225144; https://doi.org/10.1117/12.2225144

[40] Pan D, Dai F, Li H. Piezoelectric energy harvester based on bi-stable hybrid symmetric laminate. Composites Science and Technology. 2015;**119**:34-45

[41] Pan D, Ma B, Dai F. Experimental investigation of broadband energy harvesting of a bi-stable composite piezoelectric plate. Smart Materials and Structures. 2017;**26**(3):035045

[42] Pan D, Li Y, Dai F. The influence of lay-up design on the performance of bi-stable piezo-electric energy harvester. Composite Structures. 2017;**161**:227-236